The EM Algorithm and Related Statistical Models

Additional Volumes in Preparation

The EM Algorithm and Related Statistical Models

edited by

Michiko Watanabe
Tokyo University
Tokyo, Japan

Kazunori Yamaguchi
Rikkyo University
Tokyo, Japan

CRC Press
Taylor & Francis Group
Boca Raton London New York

CRC Press is an imprint of the
Taylor & Francis Group, an **informa** business

CRC Press
Taylor & Francis Group
6000 Broken Sound Parkway NW, Suite 300
Boca Raton, FL 33487-2742

First issued in paperback 2019

© 2004 by Taylor & Francis Group, LLC
CRC Press is an imprint of Taylor & Francis Group, an Informa business

No claim to original U.S. Government works

ISBN-13: 978-0-8247-4701-5 (hbk)
ISBN-13: 978-0-367-39493-6 (pbk)

Library of Congress Cataloging-in-Publication Data
A catalog record for this book is available from the Library of Congress.

Visit the Taylor & Francis Web site at
http://www.taylorandfrancis.com

and the CRC Press Web site at
http://www.crcpress.com

Preface

This book provides a comprehensive description of a method of constructing a statistical model when only incomplete data are available, and then proposes specific estimation algorithms for solving various individual incomplete data problems.

In spite of the demand for utilizing data increasingly obtained from recently compiled socioeconomic databases, the development of statistical theories is slow in coming when the available data are, for the most part, incomplete by nature. It is true that independent study results based on an awareness of individual problems have been reported; however, this is the first complete work discussing the problems in individual fields from a broader and unified perspective. In addition to addressing these very issues, the book offers a summary of prior findings and software resources.

The first chapter presents a general discussion of the problems of incomplete data along with the modeling of generation mechanisms, and provides conditions and examples of estimation based on obtained data ignoring valid generation mechanisms, and situations where the generation mechanisms cannot be ignored. Chapter 2 presents the realistic problems of the processing of missing values commonly seen in multidimensional data and the various methods to cope with them, including comparison of efficiencies in these methods. In this chapter, the maximum likelihood method is presented as one method and the EM algorithm for this purpose is introduced. Chapter 3 provides a general theory on the EM algorithm, the characteristics of each type, and examples of specific algorithm derivations according to respective statistical models. Extending the discussion of the problems of incomplete data, in Chapters 4 and 5 the application and utilization methods with the EM algorithm for the estimation issues of robust models or latent variable models are comprehensively described.

Chapter 6 presents descriptions and specific application examples of the recent improvement in the EM algorithm, including the PIEM algorithm. In Chapter 7, the acceleration methods of the EM algorithm are provided through an explanation of the convergence speed of the EM algorithm, centering on the comparison of quasi-Newton methods. In Chapter 8, the interpretation of the EM algorithm from the standpoint of information geometry and its broad application to neural network models are explained in detail. Chapter 9 discusses the relationship of the EM algorithm with the Markov Chain Monte Carlo method, which has been receiving attention since 1990 as a standard method for the Bayes estimation, wherein the data augmentation algorithm and the Gibbs sampling algorithm are described in consistent form.

Additionally, the appendixes provide a description of SOLAS™ software, which is used for the processing of the EM algorithm with incomplete data, and the ℓ_{EM} software used for general analysis of the latent structure models of categorical data with the EM algorithm.

This book will be useful for postgraduate students, statistical researchers, and practitioners.

Michiko Watanabe
Kazunori Yamaguchi

Contents

Contributors

Zhi Geng, Ph.D.   School of Mathematical Sciences, Peking University, Beijing, China

Shiro Ikeda, Dr. Eng.   The Institute of Statistical Mathematics, Tokyo, Japan

Masahiro Kuroda, Ph.D.   Department of Socio-Information, Okayama University of Science, Okayama, Japan

Mihoko Minami, Ph.D.   Department of Fundamental Statistical Theory, The Institute of Statistical Mathematics, Tokyo, Japan

Noboru Murata, Ph.D.   Department of Electrical Engineering and Bioscience, Waseda University, Tokyo, Japan

Michiko Watanabe, Dr. Sci.   Faculty of Economics, Toyo University, Tokyo, Japan

Kazunori Yamaguchi, Dr. Sci.   Department of Industrial Relations, Rikkyo University, Tokyo, Japan

1

Incomplete Data and the Generation Mechanisms

Michiko Watanabe
Toyo University, Tokyo, Japan

1 INCOMPLETE DATA PROBLEMS

In many cases of actual data analysis in various fields of applications, the data subject to the analysis are not acquired as initially planned. Data that have not been obtained in complete form as intended are called incomplete data. The incompleteness of data may take various forms. For example, part of the information may be observed, as in cases where the actual observation being sought may not be completely observed but is known to be greater than a certain value. On the contrary, no information may be obtained at all. A typical example of the former is censoring in survival time data, whereas the latter is treated as missing data.

The enhancements in software tools for statistical analysis for personal computers in recent years have led to an environment in which anyone can apply methods of statistical analysis to the process of data processing relatively easily. However, many statistical software still require that there are no missing values in the data subject to analysis, i.e., the completeness of data is a prerequisite for application. Even those that have missing value processing as an optional function are limited to processing at simple levels, such as excluding all incomplete observations, or substituting a missing value with the mean value. Additionally, the preparedness of statistical software for data incompleteness depends on the method of analysis applied, as exemplified for survival analysis; many statistical software packages are capable of analyzing censored data as well.

However, as stated before, the occurrence of missing data is an inevitable problem when collecting data in practice. In particular, as the number of items subject to survey or experiments increases, the number of so-called complete observations (cases in which the data is recorded with respect to all items) decreases. If you exclude all incomplete observations and then simply conduct statistical analysis only with respect to complete observations, the estimates obtained as a result will become less efficient and unexpected biases will arise, severely undermining the reliability of the analysis results. In other words, it is generally vital to exercise more caution with respect to incomplete data, in light of the reliability of analysis results. This chapter reviews the mechanism by which missing values generate, which should be considered first in regard to the "missing" problem, which is the most common among all incomplete data.

How should missing values be processed, and how should the results be interpreted? When dealing with these questions, it is important to understand why the values are missing in the first place. Of course, there are cases in which there is no information as to why the values are missing. In such cases, it is necessary to employ a method of analysis that is suitable for cases that lack such information. If the mechanism by which missing values arise is known, the top priority is to determine whether the mechanism needs to be taken into account or whether it could be ignored when conducting an analysis.

Consider an ordinary random sample survey as a simple case in which the mechanism by which missing values generate can be ignored when conducting an analysis. In this case, the variables to be considered are those subject to the survey items and a variable defining the sampling frame. Assuming that the data on these variables in the population are complete data, the sample data can be regarded as incomplete data in which all survey variables relating to the subjects of the survey that have not been included in the sample are missing. The only variable that is observed completely is the variable that defines the sampling frame.

If sampling is done randomly, there is no need to incorporate those that have not been observed into the analysis, and it is permissible to conduct an analysis targeting the sample data only. In other words, the mechanism by which the missing values arise (= the sampling mechanism) can be referred to as an "ignorable" mechanism upon analysis, because the missing values arise without any relation to the value of the variables that might have been observed.

On the other hand, if you are observing the time taken for a certain phenomenon (e.g., death, failure, or job transfer) to occur, you might know that the phenomenon has not occurred up until a certain point of time, but not the precise time at which it occurred; such data are called censored data. As the censored data are left with some information, that the phenomenon occurred after a certain period, it is risky to completely ignore them in the analysis as it might lead to biased results. In short, the mechanism by which missing values arise that derives censored data is a "nonignorable" mechanism.

For another example of a nonignorable mechanism, in a clinical study on clinical effects conducted over a certain period targeting outpatients, the data on patients who stopped seeing a doctor before the end of the period cannot simply be excluded as missing data. This is because the fact that the patient stopped seeing a doctor might contain important information relating to the effect of the treatment, e.g., the patient might have stopped seeing the doctor voluntarily because the patient quickly recovered, or might have switched to another hospital due to worsening symptoms. It is also necessary to exercise caution when handling missing data if the measuring equipment in use cannot indicate values above (or below) a certain threshold due to its accuracy, and uncollected data if the data are either adopted or rejected depending on the size of the variables.

In this manner, it is important to determine whether the mechanism by which missing values arise can be ignored when processing the missing values. The next section shows a statistical model of such a mechanism, and then strictly defines "missing at random" and "observed at random" based on this model and explains when the mechanism can be ignored, i.e., sets forth the conditions for cases in which it is acceptable to exclude the incomplete observations including missing values and cases in which such crude exclusions lead to inappropriate analysis results.

2 GENERATION MECHANISMS BY WHICH MISSING DATA ARISE

Generally, the fact that a value is missing might in itself constitute information on the value of the variables that would otherwise have been observed or the value of other variables. Therefore, the mechanism by which missing values arise need to be considered when processing incomplete data including missing values. Thus, incomplete data including missing values require the modeling of a mechanism by which missing values arise, and the incorporation of the nature of missing values as part of the data in the analysis.

Now, suppose the multivariate data is the observed value of a random variable vector $X = (X_1, X_1, \ldots, X_p)'$ following a multivariate density function $f(x; \theta)$. The objective of analysis for the time being is to estimate a parameter θ that defines the function f. Now, introduce, a new random variable vector $M = (M_1, M_2, \ldots, M_p)'$ corresponding to observed variable vector X. M is an index vector indicating whether the elements X are missing or observed: that is, X_i is observed when $M_i = 1$, and X_i is not observed (i.e., missing) when $M_i = 0$. In other words, the observed value m of missing index vector M shows the missingness pattern of observed data.

Modeling of a mechanism by which missing values arise concretely defines the conditional probability $g(m; x, \phi)$ of a certain observed value m of M, given the observed value x of X. Here, ϕ is a parameter that defines the mechanism.

If the output variable through the mechanism by which missing data arise is represented by random variable $V = (V_1, V_2, \ldots, V_p)'$, each element of which is defined as $V_i = x_i$ when $m_i = 1$ and $V_i = *$ when $m_i = 0$, the data acquired is v, which is the observed value of V. This also means that a missingness pattern m and $x' = (x'^{(0)}, x'^{(1)})$ corresponding thereto are realized, where $x^{(0)}$ and $x^{(1)}$ indicate the vectors of the missing part and the observed part, respectively, in X corresponding to m. Estimation of θ should be done based on such v in a strict sense.

If you ignore the mechanism by which missing values arise $g(m; x, \phi)$, you assume that the observed part $x^{(1)}$ is an observation from the marginal density

$$\int f(x)dx_0. \tag{1}$$

On the other hand, if you take into account the mechanism by which missing values arise, it is actually from the following density function

$$\int \{f(x;\theta)g(m|x; \phi)/\int \{f(x;\theta)g(m|x; \phi)dx\}dx_0 \tag{2}$$

The mix-up of the two densities in formulae (1) and (2) will give rise to a nonignorable bias in the estimation of θ depending on $g(m; x, \phi)$.

For example, suppose there is a sample of size n for a variable, and only values that are larger than the population mean are observed. Assuming that θ is the population mean and $\phi = \theta$, the mechanism by which only values that are larger than the population mean are actually observed is represented by

$$g(m|x;\phi) = \prod_i \delta(\gamma(x_i - \phi) - m_i),$$

$$\gamma(a) = \begin{cases} 1 & (a \geq 0) \\ 0 & (a < 0) \end{cases},$$

where

$\delta(a) = 1$, if $a = 0$, 0, otherwise

$$\delta(a) = \begin{cases} 1 & \text{if } a = 0 \\ 0 & \text{otherwise} \end{cases}$$

In this case, if θ is estimated based on density function (1) ignoring $g(m; x, \phi)$, it is obvious that the estimation will be positively skewed.

Generally, the mechanism by which missing values arise cannot be ignored. However, if certain conditions are met by the mechanism, Formulae (1) and (2) become equal such that the mechanism becomes ignorable. In this regard, Rubin (1976) summarized the characteristics of the mechanism as follows.

1. Missing at random (MAR). MAR indicates that given the condition that the observed part $x^{(1)}$ is fixed at any specific value, for any unobserved part $x^{(0)}$, the mechanism by which missing values arise becomes the constant, that is,

    $$g(m|x; \phi) = c(bx^{(1)}|\phi) \text{ for any } x^{(0)}$$

    In other words, MAR means that the missingness pattern m and the unobserved part $x^{(0)}$ are conditionally independent given the observed part $x^{(1)}$.

2. Observed at random (OAR). OAR indicates that given the condition that the unobserved part $x^{(0)}$ is fixed at any specific value, for any observed part $x^{(1)}$, the mechanism by which missing values arise becomes the constant, that is,

    $$g(m|x; \phi) = c(bx^{(0)}|\phi) \text{ for any } x^{(1)}$$

3. ϕ is distinct from θ. This indicates that the joint parameter space of ϕ and θ factorizes into each parameter space, and the prior distribution of ϕ is independent of that of θ.

In regard to the above description, Rubin (1976) gives the following theorems concerning the condition that the mechanism that leads to the missingness can be ignorable.

Theorem 1. In case that we infer the objective parameter θ based on the sampling distribution for the incomplete data that have missing values, we can ignore the missingness mechanism $g(m|x; \phi)$ under the condition that both MAR and OAR are met, which is called missing completely at random (MCAR).

In other words, MCAR means that the missingness pattern m and the objective variable x are unconditionally independent such that the deletion of all incomplete observations that have missing values, which is one of the simplest means to process incomplete data, leads no inference bias.

Theorem 2. In case that we infer the objective parameter θ based on the likelihood for the incomplete data, we can ignore the missingness mechanism $g(m|x; \phi)$ under the condition that MAR and the distinctness of ϕ from θ are satisfied.

Theorem 3. In case that we infer the objective parameter θ via a Bayesian method for the incomplete data, we can ignore the missingness mechanism $g(m|x; \phi)$ under the condition MAR and the distinctness of ϕ from θ are satisfied.

In regard to the results given by Rubin (1976) above, it should be emphasized that even if the missingness mechanism is not necessarily missing completely at random, the missingness mechanism can be ignored depending on the type of analysis provided that the weaker condition "missing at random" is satisfied. The EM algorithm, which is the main line in this book, has been widely applied under this MAR situation in making likelihood or Bayesian estimations.

Example 1.1

Suppose the data record the income and expenditure of each household. The objective here is to estimate the effect of income on expenditure, assuming that part of the income data is not observed. If the possibility of values in the income to go missing is the same for all households regardless of the values of income and expenditure, this case is said to be MAR and at the same time OAR. In other words, it is MCAR.

On the other hand, if the possibility of values in the expenditure to go missing is affected by the income of the household and if the possibility of values in the expenditure to go missing is not affected by the values of the expenditure itself with respect to a household with the same income values, this case is said to be MAR but not OAR.

In the latter case in Example 1.1 in which the missingness mechanism is not MCAR but MAR, in an analysis based on a regression model where the dependent variable is expenditure and the independent variable is income, the mechanism by which missing values arise in expenditure can be ignored according to Rubin's results. Of note, Heitjian and Basu (1996) have explained in detail the difference between MCAR and MAR.

The EM algorithm described in the article by Watanabe and Yamaguchi (this volume) and the subsequent chapters is an algorithm that derives ML estimates of the parameters in a model based on incomplete data including missing values. As a sufficient condition for these methods to be adequate, the mechanism by which missing values arise must be ignorable, for which the results of Rubin (1976) referred to here are important. Put differently, in the case of missing completely at random, the mechanism can be ignored without any problem. Even if that is not the case, the mechanism can be ignored provided that missing at random is satisfied and there is no relationship between the parameters that need to be estimated and the parameters determining the mechanism.

A concept based on the extension of these results is "coarsening at random." Heitjan and Rubin (1991) defined "coarsening at random" and obtained results that extend the results of "missing at random" of Rubin (1976) into "coursing at random." Furthermore, Heitjan (1993, 1994) showed specific examples. As pointed out by McLachlan and Krishnan (1997), in actual analysis, the variables subject to analysis are concerned with whether they are observed or not, meaning that caution must be exercised as there are quite a few cases in which the mechanism by which missing values arise is nonignorable. Little and Rubin (1987) referred to the method of analysis applicable to such cases in their Chapter 11.

In the case of conducting a statistical test on the equivalency of two or more populations based on data including missing values, the method based on the Permutation Test theory can be used. For example, Pesarin (1999) suggested a test method for cases in which the values are missing at random as well and cases in which the values are not missing at random. This is discussed in detail by Pesarin (1999) in his Chapter 9.

REFERENCES

1. Heitjan, D. F. (1993). Ignorability and coarse data: some biomedical examples. *Biometrics* 49:1099–1109.
2. Heitjan, D. F. (1994). Ignorability in general incomplete data models. *Biometrika* 8:701–708.
3. Heitjan, D. F., Basu, S. (1996). Distinguishing "missing at random" and "missing completely at random." *Am. Stat.* 50:207–213.
4. Heitjan, D. F., Rubin, D. B. (1991). Ignorability and coarse data. *Ann. Stat.* 19:2244–2253.
5. Little, R. J. A., Rubin, D. B. (1987). *Statistical Analysis with Missing Data.* New York: Wiley.
6. McLachlan, G. J., Krishnan, T. (1997). *The EM Algorithm and Extensions.* New York: Wiley.
7. Pesarin, F. (1999). *Permutation Testing of Multidimensional Hypotheses by Nonparametric Combination of Dependent Tests.* Padova, Italy: Cleup Editrice.
8. Rubin, D. B. (1976). Inference and missing data. *Biometrika* 63:581–592.

2
Incomplete Data and the EM Algorithm

Michiko Watanabe
Toyo University, Tokyo, Japan

Kazunori Yamaguchi
Rikkyo University, Tokyo, Japan

1 ANALYSIS OF MULTIVARIATE DATA WITH MISSING VALUES

In many cases of multivariate data analysis, the values of the mean vector and the variance–covariance matrix or the correlation matrix are calculated first. Various methods of multivariate analysis are actually implemented based on these values. Broadly speaking, there are four methods of processing the missing values, as described below (Little and Rubin, 1987).

1. Exclude all incomplete observations including missing values. The sample size will be smaller than originally planned, however. Create a data matrix consisting of only complete observations and execute the ordinal estimation procedures.
2. Exclude only the missing parts and estimate the mean vector based on the arithmetic mean of the observed values that exist with respect to each variable. For the variance–covariance matrix and the correlation matrix, perform estimation by using all the observed values of each pair of variables, if any.

3. Estimate the missing values to create a pseudocomplete data matrix while retaining the original sample size and execute normal estimation procedures.
4. Assuming an appropriate statistical model, estimate the maximum likelihood of necessary parameters based on the marginal likelihood with respect to all observed values that have been acquired.

Method 1 for processing missing values is normally taken by many statistical analysis software. Despite the massive loss of information volume, it is probably the most frequently used method today because of easy processing. However, this method should not be applied crudely because the analysis results might be grossly biased depending on the mechanism by which missing values arise, as shown in the previous chapter.

Method 2 may appear to be effective because the data volume used in calculations is larger than that in Method 1. However, the results of a simulation experiment on multiple regression analysis conducted by Haitovski (1968) show that it is less accurate than Method 1. Moreover, Method 2 does not guarantee the positive definiteness of the estimates of the variance–covariance matrix or the correlation matrix. Accordingly, the danger is that they might not be usable in subsequent analyses. Although Method 2 is also implemented in some statistical analysis software as an optional way of calculating correlation matrices, etc., one must be cautious when applying this method because of the aforementioned reasons.

According to some research reports, the method based on either Method 3 or 4 is more effective than Methods 1 and 2. The next section discusses the methods of estimating the missing value itself in concrete terms, and shows how the efficiency of the estimation of target parameters varies depending on the way in which the missing values are processed.

2 ESTIMATION OF MISSING VALUES

The method of processing missing values referred to in Method 3 in the previous section makes the incomplete data "complete" expediently by substituting the missing value itself with an estimate. This method is expected to inflict a smaller loss of information than Method 1, which excludes all incomplete observations. It also allows normal statistical analysis methods to be applied to a complete data matrix in which the missing value has been complemented, making the analysis itself easier.

There are two typical methods of estimating missing values, as follows.

Mean value imputation: Substitute the missing value with the mean of the observed data with respect to each variable corresponding to the missing value.

Regression: Based on a multiple regression model assuming that the variable corresponding to the missing value is a dependent variable and the variable portion in which the data is observed is a group of independent variables, figure out the estimate of the regression coefficient based on the data matrix of the portion in which the data is completely observed with respect to all missing variables. Estimate the missing values in incomplete observations based on a regression formula using the estimated regression coefficient.

In survey statistics such as official statistics, the alternatives include hot deck, which finds complete observations in which a specified observed data portion is similar to incomplete observations in the same data and complements the missing value by using the value of the variable corresponding to the observations, and cold deck. Both of these methods are used expediently. Recently, multiple imputation advocated by Rubin (1987) has been attracting a great deal of attention for taking sample variation into account.

Among the methods of processing missing values (Method 3), this section reviews mean value imputation and partial regression, in addition to the exclusion method (the exclusion of all incomplete observations as referred to in Method 1 in the previous section), and compares these methods based on Monte Carlo experiments in light of the estimated parameter efficiency.

2.1 Comparison of Efficiency

Suppose an incomplete data matrix of $N \times p$ is X. Here, N indicates the sample size (the total number of observations), and p represents the number of variables.

N observations can be divided into n complete observations that do not include missing values (downsized complete data portion) and m incomplete observations that include missing values (incomplete data portion). Here, data matrix X consists of n complete observations in the top row and m incomplete observations in the bottom row.

In this context, the exclusion Method 1 uses only the downsized complete data portion, whereas mean value imputation estimates the missing values in the variables corresponding to the incomplete data portion based on the sample mean of the variables in the downsized complete data portion. On the other hand, regression involves the estimation of the regression model of missing variables with respect to observed variables based on the downsized complete data portion according to the missing pattern of each observation in the incomplete data portion, and the estimation of the missing values in the incomplete data according to the regression formula. In this case, the computational complexity under partial regression depends on the total number of missing patterns in the incomplete data portion.

The following is a comparison of the estimator efficiency with respect to the sample mean vector and the sample variance–covariance matrix resulting from the exclusion method, and the sample mean vector and the sample variance–covariance matrix calculated based on a pseudo-complete data matrix that has been "completed" by mean value imputation and regression.

The criterion for comparison is the relative efficiency (R.E.) with the estimator of the mean vector based on the complete data assuming that no values were missing $\tilde{\mu}$ and the estimator of variance–covariance matrix $\tilde{\Sigma}$.

$$\text{R.E.}(\hat{\mu}|\tilde{\mu}) = \frac{E[\text{tr}(\tilde{\mu} - \mu)(\tilde{\mu} - \mu)']}{E[\text{tr}(\hat{\mu} - \mu)(\hat{\mu} - \mu)']} \tag{1}$$

$$\text{R.E.}(\widehat{\Sigma}|\tilde{\Sigma}) = \frac{E[\text{tr}(\tilde{\Sigma} - \Sigma)(\tilde{\Sigma} - \Sigma)']}{E[\text{tr}(\widehat{\Sigma} - \Sigma)(\widehat{\Sigma} - \Sigma)']} \tag{2}$$

where $\hat{\mu}$ and $\widehat{\Sigma}$ represent the sample mean vector and the sample variance–covariance matrix, respectively, which can be acquired by the respective methods of processing missing values.

Figs. 1 and 2 show the results of simulation experiments conducted 1000 times based on multivariate normal random numbers. The horizontal axis depicts an index of the correlation matrix of the four population distributions used, which is $\gamma = (\lambda - 1)/(p - 1)$ (λ is the maximum eigenvalue of the correlation matrix). The vertical axis represents the relative efficiency.

According to Fig. 1, the two methods of estimating the missing values are always more effective than the simple exclusion method as far as the estimation of mean vector μ is concerned. On the other hand, the order of superiority of the three processing methods in terms of the estimation of

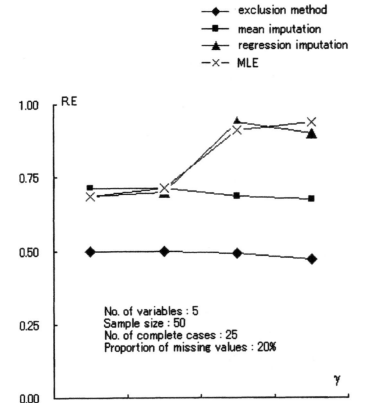

Figure 1 Relative efficiency of estimates of mean vector.

variance–covariance matrix Σ depends on the number of variables and the extent to which the variables are correlated to each other, varying from case to case (refer to Fig. 2). If the number of variables is small, the exclusion method can perform a sufficiently efficient estimation. However, if there are many variables, its estimation efficiency is inferior to the other two processing methods of estimating the missing values. Put differently, the relative efficiency of the methods of estimating the missing values varies with the extent to which the variables are correlated with each other. This means that the efficiency of mean value imputation is the highest when

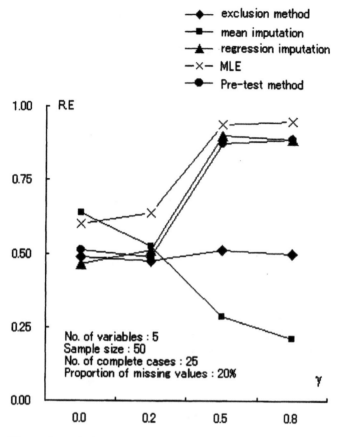

Figure 2 Relative efficiency of estimates of covariance matrix.

there is a weak correlation between variables, and, conversely, the efficiency of the regression is high when there is a strong correlation between variables. Therefore, it is desirable to distinguish the two methods of estimating the missing values appropriately when using them, based on an understanding of the extent to which the variables are correlated with each other in advance. If this understanding is vague, it would be appropriate to conduct a preliminary test on the significance of the partial regression formula estimated based on the downsized complete data portion, and use either mean value imputation or partial regression based on the results.

Fig. 2 shows the relative efficiency of the methods of estimating the missing values in cases where the preliminary test is included. Here, the optimal solution based on minimax criteria is adopted as the critical value in the test. The relative efficiency of the preliminary test method is uniformly higher than the exclusion method. This shows that the loss resulting from choosing the wrong method is rectified by conducting a preliminary test.

For partial regression, Frane (1976) made improvements by incorporating techniques for selecting variables in a regression model. Also, Beal and Little (1975) advocated the iteration method, which involves the further revision of the partial regression formula by using the mean vector and the variance–covariance matrix estimated by partial regression, and the iteration of the two processes that update the estimation of missing values thereby until they converge. The results indicate that they can be regarded as more or less the same as the results of maximum likelihood estimation when the data is compliant with multivariate normal distribution. Fig. 2 also shows the relative efficiency of the maximum likelihood estimator based on the incomplete data matrix itself assuming a multivariate normal model. The maximum likelihood method shows a similar pattern to the preliminary test method with respect to the correlation between variables, and its efficiency is higher than the preliminary test method.

The next section explains the algorithm for deriving the maximum likelihood estimate from incomplete data.

3 MAXIMUM LIKELIHOOD ESTIMATION BASED ON INCOMPLETE DATA

If it is possible to assume the distribution with which the data is compliant, the maximum likelihood estimate of the parameter can be found based on the data matrix even in cases where there are missing values. Maximum likelihood methods assuming multivariate normal distribution have been discussed for a long time, including Anderson (1957) and Trawinski and Bargmann (1964). However, the missingness pattern in the incomplete data matrix must show a special aspect called "nest" in order for the mean and the maximum likelihood estimator of the variance-covariance matrix to be represented in terms of a positive formula (Rubin, 1974). Nest refers to an aspect of missingness in an incomplete data matrix such that in cases where the ith variable $(2 \leq i \leq p)$ is observed based on the proper rearrangement of the variables and observations, the data of the $(i - 1)$th variable is always observed (the opposite may not be true).

Iterative calculation by computer is required to determine the maximum likelihood estimate of μ and Σ without placing any restrictions on the missingness patterns. Hartley and Hocking (1971) advocated an algorithm for directly deriving a likelihood equation from an incomplete data matrix and determining a standard solution based on the scoring method. On the other hand, Orchard and Woodbury (1972), Beal and Little (1975), and Sundburg (1976) derived a method of finding the maximum-likelihood solution based on an algorithm which later became generally known as the EM algorithm by Dempster et al. (1977).

3.1 EM Algorithm

The scoring method constructs the likelihood equation for μ and Σ based on the marginal likelihood of all observed data. However, there is a way to maximize the marginal likelihood relating to observed data without using this equation, that is, the EM algorithm. Generally speaking, the likelihood equation of incomplete data based on the distribution in a exponential distribution family can be represented as follows, including but not limited to multivariate normal distribution.

Generally speaking, the likelihood equation of incomplete data based on the distribution in an exponential distribution family can be represented as follows, including but not limited to multivariate normal distribution,

$$E(t;\theta) = E(t|\{x_i^{(0)}\}; \theta) \tag{3}$$

where θ is the parameter to be estimated, and its sufficient statistic is t. Furthermore, $\{x_i^{(0)}\}$ represents the observed data portion with respect to N observations.

The EM algorithm divides the process of solving Eq. (3) into two steps. The first step is to find the conditional expected value on the left side of the equation. The second is to solve the equation in concrete terms using the conditional expected value that has been found.

Now we consider a sample $\{x_i\}(i = 1, \ldots, N)$ from p-dimensional normal distribution. μ and Σ are parameters to be estimated in the multivariate normal distribution model. Its sufficient statistics are a $1/2p(p + 3) \times 1$ column vector consisting of the sum of p variates in the complete data matrix X, and $1/2p(p + 1)$ values of the sum of squares and cross products of variables. The EM algorithm is a method of finding estimated vectors $\hat{\mu}$ and $\hat{\Sigma}$ that satisfy Eq. (3) via the estimation of t, and consists of two steps. The calculation methods of the two steps are described in further detail in Chap. 3.

The drawback of the EM algorithm is allegedly the larger number of iterations generally required to reach convergence than the aforementioned scoring method. However, it can avoid so-called inverse matrix calculations. Also, programming is easy because a simple algorithm normally applied to complete data can be used in the maximization step. As the merits and demerits of the scoring method and the EM algorithm are demonstrated on opposite standpoints, the choice must be made on a case-by-case basis rather than in general terms (refer to Chap. 7).

As for the initial value, it would be reasonable to use the sample mean and the variance–covariance matrix determined by the complete data portion in both methods.

The EM algorithm will generally be discussed in detail in Chap. 3 and the subsequent chapters. The merit of the EM algorithm concept is that the calculations following the calculation of the conditional expected value constitute estimations in cases where there are no missing values. With respect to many problems, the calculation methods are already known, and the calculations can be performed easily without having to worry about the missingness pattern. Another great merit is that no major inverse matrix problem has to be solved. As substantial computational complexity is associated with solving major inverse matrix problems, it will have the edge in computation time as well. Chap. 7 compares the EM algorithm with other optimization algorithms in terms of convergence speed, with reference to estimation problems in linear mixed models and normal mixture distributions as examples.

4 SUMMARY

This chapter described a few of the key methods of executing multivariate analysis based on missingness or a data matrix including missing values, in light of efficient methods of processing the missing values.

As for the method of processing missing values, it is important to investigate the missingness pattern with respect to each cause, as to why the missingness or the deficiency occurred. It can be complemented considerably by actively analyzing and modeling the mechanism by which missing values arise.

If such mechanism does not particularly exist or can be ignored, the randomness of the occurrence of missing values must be scrutinized.

In cases where the mechanism by which missing values arise is ignorable, the missing value itself is normally estimated to make the data matrix "complete." Typical methods are mean value imputation and regression, which have opposite merits and demerits depending on the nature of the data matrix. Accordingly, Watanabe (1982) advocated choosing between mean value imputation and partial regression depending on the results of the preliminary test on a priori information, and exemplified the rationality of compensating the merits and demerits. This chapter also described that maximum likelihood estimation of the mean vector and the variance–covariance matrix used in the analysis process can directly be performed even if it includes missing values, and reviewed the scoring method and EM algorithm as algorithms for that purpose.

In the former case, estimation of the missing value itself firstly involves the use of the complete data portion, meaning that the accuracy will deteriorate as the number of complete data observations decreases. Under the maximum likelihood method, it is possible to at least estimate the population even if there is no complete data. From a different perspective, caution must be exercised in regard to the fact that the distribution needs to be known in advance in the maximum likelihood method, in contrast to the method of estimating the missing values, which is robust against the distribution of the population generating the data matrix.

This chapter focused on how to process missing and deficient values in multidimensional data. Censored, truncated, grouped data, etc., are also incomplete data that occur on an everyday basis, and are discussed in Chap. 3.

REFERENCES

1. Anderson, T. W. (1957). Maximum likelihood estimates for a multivariate normal distribution when some observations are missing. *J. Am. Stat. Assoc.* 52:200–203.
2. Beal, E. M., Lttle, R. J. A. (1975). Missing values in multivariate analysis. *J. R. Stat. Soc., B* 37:129–145.
3. Dempster, A. P., Laird, N. M., Rubin, D. B. (1977). Maximum likelihood from incomplete data via the EM algorithm. *J. R. Stat. Soc., B* 39:1–22.
4. Frane, J. W. (1976). Some simple procedures for handling missing data in multivariate analysis. *Psychometrika* 41:409–415.

5. Haitovski, Y. (1968). Missing data in regression analysis. *J. R. Stat. Soc., B* 30:67–82.

6. Hartley, H. O., Hocking, R. R. (1971). The analysis of incomplete data. *Biometrics* 27:783–808.

7. Little, R. J. A., Rubin, D. B. (1987). *Statistical Analysis with Missing Data.* New York: Wiley.

8. Orchard, T., Woodbury, M. A. (1972). A missing information principle: theory and applications. In: *Proceedings of the 6th Berkeley Symposium on Mathematical Statistics and Probability.* California: University California Press, pp. 697–715.

9. Rubin, D. B. (1974). Characterizing the estimation of parameters in incomplete-data problems. *J. Am. Stat. Assoc.* 69:467–474.

10. Rubin, D. B. (1987). *Multiple Imputation for Nonresponse in Surveys.* New York: Wiley.

11. Sundberg, R. (1976). An iterative method for solution of the likelihood equations for incomplete data from exponential families. *Commun. Stat. Simul. Comput.* 5:55–64.

12. Trawinski, I. M., Bargmann, R. (1964). Maximum likelihood estimation with incomplete multivariate data. *Ann. Math. Stat.* 35:647–657.

13. Watanabe, M. (1982). Analysis of multivariate data with missing values. *Math. Sci.* 230:41–48 (in Japanese).

3
Statistical Models and the EM Algorithm

Kazunori Yamaguchi
Rikkyo University, Tokyo, Japan

Michiko Watanabe
Toyo University, Tokyo, Japan

1 INTRODUCTION

In 1977, Dempster, Laird, and Rubin (DLR) (Dempster et al., 1997) advocated a unified algorithm, called the EM algorithm, for deriving maximum likelihood estimates from incomplete data and showed its wide scope of application to various statistical models for the first time. Although many essays critical of its problems, such as the convergence in the algorithm, were released in the early days, numerous essays have been published over the past 20 years, hammering out new methodologies using the EM algorithm in almost all fields in which statistical analysis is required, including engineering, medical science, sociology, and business administration. According to a survey conducted by Meng and Pedlow (1992), at least 1700 essays exist on more than 1000 subjects. Moreover, Meng (1997) pointed out that more than 1000 essays were published in approximately 300 types of magazines in 1991 alone (of which statistics journals accounted for only 15%). These facts clearly indicate that the EM algorithm has already become a multipurpose tool for building a method of statistical analysis based on likelihood, surpassing the Newton–Raphson method and other substitution methods.

Why has the EM algorithm gained ground to such an extent? It is often said that the EM algorithm's beauty lies in the algorithm's simplicity and stability. For these reasons, the authors have used the EM algorithm in studying the latent class model, the proportional hazard model, the robust factor analysis model, the covariance structure model, and the Tobit model, and in deriving the maximum likelihood solution in those models since Watanabe (1989) introduced in a Japanese magazine the maximum likelihood method based on multidimensional data, including missing values with reference to DLR. In the early days, questions about studies using the EM algorithm at academic conferences, etc., were always the same: "Why does it specifically have to be the EM algorithm?" "What is the point of using the EM algorithm despite its alleged slow convergence in an environment where optimization software is becoming more sophisticated?" "How many iterations were required to reach convergence?"

First, in response, an algorithm for performing maximum likelihood estimation based on multidimensional data, including missing values, was programmed specifically based on the Newton–Raphson method and the EM algorithm, and an experiment was conducted to compare the two in terms of the number of iterations required to reach convergence, central processing unit (CPU) time, etc., using the same data. The results showed that the EM algorithm was able to determine a convergence value in all cases, whereas the application of the simple Newton–Raphson method failed to achieve convergence in most cases. This is probably why the EM algorithm is referred to as stable. In addition, the results indicated that when the two converged on the same maximum likelihood solution, the EM algorithm was faster in terms of the CPU time taken to reach convergence on the whole: although fewer iterations were required by the Newton–Raphson method, the CPU time required per iteration was overwhelmingly shorter for the EM algorithm. The EM algorithm is generally claimed to suffer from slow convergence, but this alleged drawback appears to be no major obstacle in practice. In fact, the simplicity of the EM algorithm seems to be much more attractive, considering the relatively high operating efficiency from the stage of formulating the likelihood to the stages of deriving and programming an algorithm in concrete terms. A number of improved versions of the EM algorithm, aimed at accelerating convergence, have been proposed since DLR. However, they failed to gain wide acceptance because they sacrificed some aspects of the simplicity and the stability of the original EM algorithm.

The applications of the EM algorithm are broad because of its flexibility in interpreting the incompleteness of data subject to analysis, and the high extensibility of the application model. The missing value problem, which constitutes the original meaning of incomplete data, truncated and censored distributions, mixture distributions, random effect models and mixture models in analyses of variance, robust analyses, latent variable models, and key statistical models, are within its scope. Moreover, it is possible to combine models within its scope to build a more complex model. However, if the problem becomes complex, simple calculations in Expectation Step (E-step) and Maximization Step (M-step) in the EM algorithm will not suffice; for example, evaluation may be required based on the random numbers generated from the model in E-step, or an iterative algorithm may have to be included as well, such as the Newton method, in M-step. The scope of its application to real-life problems has expanded in recent years, including applications to medical image processing, pattern recognition, and neural network, as well as in areas where enormous computational complexity is required. In this regard, the acceleration of the EM algorithm will have great significance in practice.

In the 1990s, many papers on the systematization of these extensions to the EM algorithm were released. Rubin (1991) ("EM and Beyond") explains the four typical algorithms based on simulation (Multiple Imputation, Data Augmentation Algorithm, Gibbs Sampler, and Sampling/ Importance Resampling Algorithm) in a unified manner based on an extended EM framework with a random number mechanism. Extensions that deserve much attention include the ECM and ECME algorithms advocated by Liu and Rubin (1994, 1995), the AECM advocated by Meng and Dyk (1997), and the accelerated EM, which does not sacrifice the simplicity and stability of the original EM. Another publication is the work of McLachlan and Krishnan (1997), "The EM Algorithm and Extension," which covers recent topics relating to them as well. The EM algorithm was featured in the magazine *Statistica Sinica* in 1995, followed by the *Statistical Methods in Medical Research* in 1997. The essays by Meng (1997), Meng and Dyk (1997), and Amari (1996) were written in their respective capacities as guest columnists and tutorial writers for the magazines.

Today, the EM algorithm is a familiar statistical tool for solving real-life problems in various fields of applications. This chapter outlines the key applications of the EM algorithm in recent years to clarify why it has to be the EM algorithm.

2 EM ALGORITHM

2.1 Introduction

Rubin (1991) regarded the EM algorithm as one of the methodologies for solving incomplete data problems sequentially based on a complete data framework. The idea on which it is based is simple, as summarized in the steps shown below, assuming that Y_{obs} is the observed data portion and Y_{mis} is the missing data portion:

1. If the problem is so difficult that the solution cannot be derived immediately just from the data at hand Y_{obs}, make the data "complete" to the extent that solving the problem is regarded as easy and formulate the problem (assuming that the missing data portion Y_{mis} exists).
2. For example, if the objective for the time being is to derive the estimate of parameter θ, which is $\hat{\theta}$, enter a provisional value into Y_{mis} to determine $\hat{\theta}$.
3. Improve Y_{mis} using $\hat{\theta}$ and enter the value into Y_{mis}.
4. Repeat the aforementioned two steps until the value of $\hat{\theta}$ converges.

In addition to the EM algorithm, methodologies based on this framework that have been attracting much attention in recent years include: (a) multiple imputation (Rubin, 1987a), (b) data augmentation (Tanner and Wong, 1987), (c) Gibbs sampler (Geman and Geman, 1984), and (d) the SIR algorithm (Rubin, 1987b).

Multiple imputation is discussed in Appendix A in relation to execution software, and the other three methodologies are explained in detail in Chapter 9 in relation to MCMC. Therefore, such extensions will not be addressed in this section. Of note, they are distinguished from the EM algorithm in terms of how the value is entered in Y_{mis} in Step 2.

2.2 Theory of DLR (1977)

Let x be complete data and let $f(x|\theta)$ be its probability density function. In addition, let y be incomplete data and let $g(y|\theta)$ be a probability density function of y. We consider two sample spaces Ω_X (a sample space for complete data) and Ω_Y (a sample space for incomplete data). We assume that there is a mapping of $y \rightarrow y(x)$ from Ω_X to Ω_Y. Then, the probability density function of y, $g(y|\theta)$, is:

$$g(y|\theta) = \int_{\Omega_Y(y)} f(x|\theta)\,dx$$

where $\Omega_Y(y)$ is a subsample space of Ω_X determined by the equation $y = y(x)$. For missing data problem, DLR assumed that (a) parameters to be estimated are independent of the missing data process, and (b) missing data are missing at random (see Chapter 1).

Let $LL_c(\theta) = \log f(x|\theta)$, which is the log likelihood function based on the complete data, and $LL(\theta) = \log g(y|\theta)$, which is the log likelihood function based on the incomplete data. The goal of the EM algorithm is to find the maximum likelihood estimate of θ, which is the point of attaining the maximum of $LL(\theta)$.

The EM algorithm approaches indirectly the problem of maximizing the log likelihood $LL(\theta)$ based on incomplete data by proceeding iteratively in terms of the log likelihood based on the complete data, $LL_c(\theta)$. Because it is unobservable, it is replaced by the conditional expectation given the observation and temporary values of parameters:

$$\theta^{(k+1)} = \arg \max_{\theta \in \Theta} E[LL_c(\theta) \mid y, \theta^{(k)}] \tag{1}$$

Eq. (1) can be divided into the E-step and the M-step as follows:

E-step: to calculate the conditional expectation of complete data log likelihood given the observation y and the kth temporary values of parameter $\theta^{(k)}$:

$$Q(\theta; \theta^{(k)}) = E[LL_c(\theta) \mid y, \theta^{(k)}] \tag{2}$$

M-step: to find $\theta^{(k+1)}$ to maximize $Q(\theta; \theta^{(k)})$, calculated in E-step:

$$Q(\theta^{(k+1)}; \theta^{(k)}) \geq Q(\theta; \theta^{(k)}) \tag{3}$$

The E-step and M-step are repeated by turns until they converge in a specified sense. In the EM algorithm:

$$LL(\theta^{(k+1)}) \geq LL(\theta^{(k)})$$

Therefore we can get the maximum likelihood estimates of θ if we select proper starting values.

If $f(x|\theta)$ has the regular exponential family form, we have a very simple characterization of the EM algorithm as follows:

E-step: to calculate the conditional expectation of the complete data, sufficient statistics t given observation y and $\theta^{(k)}$ derive:

$$t^{(k+1)} = E[t(x) \mid y, \theta^{(k)}] \tag{4}$$

M-step: to get $\theta_{(k+1)}$ to solve the following equations:

$$E[t(x)\theta] = t^{(k+1)} \tag{5}$$

DRL pointed out that Eq. (5) is the familiar form of the likelihood equations for maximum likelihood estimation given data from a regular exponential family.

2.3 An Example: Multivariate Normal Data

We have a sample $(y_1, \ldots, y_N)$ with missing values from $p \times 1$ random vector Y, which follows the multivariate normal distribution with mean vector $\mu = (\mu_1, \ldots, \mu_p)'$ and covariance matrix $\Sigma = (\sigma_{jk})$.

We partition $y_i' = (y_{0i}', y_{1i}')$ corresponding to missing patterns, where y_{1i} and y_{0i}' are sets of observed values and missing values, respectively. The mean vector and covariance matrix are also partitioned corresponding to $y_i' = (y_{0i}', y_{1i}')$:

$$\mu = \begin{pmatrix} \mu_{0i} \\ \mu_{1i} \end{pmatrix}, \quad \Sigma = \begin{pmatrix} \Sigma_{00i} & \Sigma_{01i} \\ \Sigma_{10i} & \Sigma_{11i} \end{pmatrix}$$

Note that only partitions depend on i.

E-step: to calculate the conditional expectation of (Y_i, Y_i, Y_i'):

$$\hat{y}_{0i}^{(K+1)} = E\left(y_{0i} | y_{1i}; \mu^{(k)}, \Sigma^{(k)}\right) \tag{6}$$

$$\hat{y}_{0i}^{(K+1)} = \mu_{0i}^{(k)} + \Sigma_{01i}^{(k)} \Sigma_{11i}^{(k)-1} y_{1i} \tag{7}$$

$$E\left(y_{0i} y_{1i}' | y_{1i}; \mu^{(k)}, \Sigma^{(k)}\right) = \hat{y}_{0i}^{(K+1)} y_{1i}'$$

$$\widehat{yy}_{0i}^{(K+1)'} = E\left(y_{0i} y_{0i}' | y_{1i}; \mu^{(k)}, \Sigma^{(k)}\right) \tag{8}$$

$$\widehat{yy}_{0i}^{(K+1)'} = \hat{y}_{0i}^{(K+1)} \hat{y}_{0i}^{(K+1)'} + \Sigma_{00}^{(k)} - \Sigma_{01i}^{(k)} \Sigma_{11i}^{(k)-1} \Sigma_{10i}^{(k)} \tag{9}$$

Then:

$$\hat{y}_i^{(k+1)} = \begin{pmatrix} \hat{y}_{0i} \\ \hat{y}_{1i} \end{pmatrix}, \quad \widehat{yy}_i^{(K+1)'} = \begin{pmatrix} \widehat{yy}_{0i}^{(K+1)'} & \hat{y}_{0i}^{(K+1)} y_{1i}' \\ y_{1i} \hat{y}_{0i}^{(K+1)'} & y_{1i} y_{1i}' \end{pmatrix}$$

M-step:

$$\mu^{(k+1)} = \frac{1}{N} \sum_{i=1}^{n} \widehat{y}_i^{(k+1)} \tag{10}$$

$$\Sigma^{(k+1)} = \frac{1}{N} \sum_{i=1}^{n} \widehat{yy}_i^{(K+1)'} - \mu^{(k+1)} \mu^{(k+1)'} \tag{11}$$

3 PROPERTIES OF THE EM ALGORITHM

This section collectively reviews the characteristics of the EM algorithm and the results that have been obtained so far in regard to convergence:

1. In the EM algorithm, the value of $LL(\theta|y)$ increases in each iteration process (to be more precise, it does not decrease). This aspect is useful for debugging on programming in real life. In addition, the EM algorithm is relatively robust against the initial value.

2. The rate of convergence of the EM algorithm is proportionate to $1-\gamma$, where γ is the maximum eigenvalue of $\hat{I}c^{-1}(\theta;y)\hat{I}_m(\theta;y)$, which is the ratio of the information matrix of complete data to the information matrix of incomplete data, and refers to the proportion of the information volume of the incomplete data in the information volume of the complete data.

3. In the EM algorithm, the output in each step is statistically significant especially if the complete data belong to an exponential distribution family. In other words, the domain and the constraints of the parameters are naturally fulfilled (refer to Sec. 2), and the solution can be determined easily by entering the constraints. There are no so-called improper solutions. For example, in cases where a variance–covariance matrix needs to be estimated based on a multidimensional data matrix with missing values, the positive definiteness of the value of Σ determined in each step is guaranteed if the initial value starts with a positive definite matrix. For the estimation of incidence, the solution can naturally be determined in the incidence domain.

4. If the log likelihood $LL(\theta|y)$ based on incomplete data y is bounded, the value of the log likelihood in the iteration process $\{LL(\theta^{(k)}|y)\}$ converges to the stationary value of $LL(\theta|y)$.

5. In general conditions, if $\theta^{(k)}$ converges, the convergence value can be proven to be either the local maximum or the saddle point of LL($\theta|y$) (Boyles, 1983; Wu, 1983). Therefore, if the likelihood function is unimodal and the first derivative of function Q defined in (Eq. (2)) is continuous with respect to $\theta^{(k)}$ and θ, the EM algorithm converges to the only local maximum (maximal value). Generally speaking, however, the likelihood function of the incomplete data is not necessarily unimodal. Therefore, it is necessary to compare the values of the log likelihood of the convergence value, starting with many initial values.

6. If the convergence of the EM algorithm is extremely slow, it implies that the likelihood function is flat toward that direction.

7. There is no need to evaluate the first and second derivatives of the log likelihood. In many cases, this aspect helps save the CPU time of each iteration compared to other maximization algorithms. Put differently, as the alleged drawback (i.e., slow convergence) is based on the number of iterations required for convergence, if one refers to the total CPU time as the yardstick, it is impossible to categorically claim that it is superior/inferior to the Newton–Raphson method, etc., in terms of the actual speed of convergence, in consideration of the high-order inverse matrix calculations avoided per iteration.

8. The alleged drawback of the EM algorithm is the lack of evaluation of the asymptotic variance of the estimate, in contrast with other iteration methods that use second partial derivatives, such as the Newton–Raphson method. However, there is a way to evaluate the asymptotic variance–covariance matrix of MLE within the EM algorithm framework without evaluating the likelihood of incomplete data or the derivatives at all. This is reviewed briefly in Sec. 4 (for the details, refer to Meng and Rubin, 1991; Louis, 1982).

4 ASYMPTOTIC COVARIANCE MATRIX VIA THE EM ALGORITHM

The EM algorithm does not generate an estimate of the asymptotic variance–covariance matrix of the estimate as its by-product. However, a number of approximation methods for the asymptotic variance–covariance matrix have been advocated, while retaining the merit of the EM algorithm [i.e., there is no need to be directly aware of the log likelihood of the incomplete data LL(θ by)]. In particular, the easiest way is to evaluate the observed information matrix numerically and find the inverse matrix.

Efron and Hinkley (1978) recommended its use on the grounds that the observed information I_y is often more appropriate than Fisher's information criteria.

Louis (1982) derived the following formula to evaluate observed information matrix $I(\theta;y)$ from the complete data frames according to the missing information principle by Orchard and Woodbury (1972):

$$I(\hat{\theta};y) = E[B_c(x;\theta)\,|\,y;\theta = \hat{\theta}] - E[S_c(x\,|\,\theta)S_c^T(x\,|\,\theta)\,|\,y;\theta]_{\theta=\hat{\theta}} \quad (12)$$

where $S_c(x;\theta)$ is the first derivative vector of the log likelihood of complete data x and $-B_c(x;\theta)$ is the second derivative matrix.

After finding $\hat{\theta}$ by the EM algorithm, the observed information can be found by evaluating Eq. (12). The asymptotic variance–covariance matrix can be derived by calculating the inverse matrix.

Meng and Rubin (1991) derived the following formula on the grounds of the asymptotic variance–covariance matrix of the incomplete data (the inverse matrix of the observed information matrix I_y) by adding the increment due to the existence of missing data to the variance–covariance matrix of complete data x:

$$I^{-1}\left(\hat{\theta};y\right) = \tilde{I}_c^{-1}(\theta;y) + \Delta V \quad (13)$$

where $\tilde{I}_c(\theta;y)$ is the conditional expected value of the information matrix of the complete data, and:

$$\Delta V = \left\{I - J\left(\hat{\theta}\right)\right\}^{-1} J\left(\hat{\theta}\right)_c (\theta;y)\tilde{I}_c^{-1}(\theta;y) \quad (14)$$

$$J\left(\hat{\theta}\right) = \tilde{I}_c^{-1}(\theta;y)\tilde{I}_m^{-1}(\theta;y) \quad (15)$$

Eq. (13) can be solved by calculating $\tilde{I}_c(\theta;y)$ and Jacobian matrix $J(\hat{\theta})$. $\tilde{I}_c(\theta;y)$ can easily be solved by a normal formula for complete data. Moreover, Jacobian matrix $J(\hat{\theta})$ can be calculated based on the output of the EM algorithm itself, as its factors are the rates of convergence with respect to each component of the parameter.

5 INCOMPLETE DATA AND THE EM ALGORITHM

The EM algorithm is applied to a wide range of incomplete data. In addition to those that are incomplete in the normal sense, such as missing, truncated, censored, and grouped data, many statistical models that were conventionally not regarded as incomplete data (such as latent structure models, mixture distribution models, robust distribution models, variate models and mixture models in the analysis of variance, and Bayesian models) are addressed by the EM algorithm by assuming expedient pseudo-complete data for the sake of calculation simplicity.

It is also possible to combine models within the scope of the EM algorithm with normal missing data problems to roll out new extensible incomplete data problems.

The EM algorithm's beauty lies in, more than anything else, its ability to develop maximum likelihood estimation methods in concrete terms that are relatively easily in regard to a wide range of models as such. The method is especially useful in cases where some kind of constraint is required among parameters to assure the identifiability of the model, as it can flexibly deal with constraints among parameters.

However, if the structure of incompleteness of the applied statistical model is complex, it will be difficult to find the maximization of the Q function—the conditional expected value of the log likelihood in M-step—by a standard closed formula. In such cases, it is necessary to apply some kind of optimization algorithm in M-step and perform iterative calculation. The GEM algorithm, which is an extension of the EM algorithm by DLR, and other extended versions of the EM algorithm may be effective in these cases.

5.1 GEM Algorithm and the Other Extensions

The GEM algorithm replaces M-step in the EM algorithm with a step to find $\theta^{(k+1)}$ that satisfies the following formula:

$$Q\left(\theta^{(k+1)} \mid \theta^{(k)}\right) \geq Q\left(\theta^{(k)} \mid \theta^{(k)}\right) \tag{16}$$

This indicates that it is not always necessary to find the maximization of the Q function in M-step, and that it is sufficient to find $\theta^{(k+1)}$ that updates it to a larger value. Therefore, in cases where maximization is sought by using the Newton–Raphson method, etc., in M-step, it is possible to stop after just one iteration. Lange (1995) advocated this method as a gradient algorithm.

Meng and Rubin (1993) adopted the ECM algorithm as a method of using a conditional maximization matrix in M-step. Also, Liu and Rubin (1995) advocated the ECME algorithm as an algorithm that switches the subject of maximization in M-step according to the components of $\theta^{(k)}$ from function to the direct log likelihood of observed data.

In cases where there is no simple solution based on a closed formula in E-step, there are a number of creative ways to handle this, including the substitution with approximate distribution (Laird, 1978) and the use of Laplace expansion (Steel, 1996). Tanner and Wong (1987) used the Monte Carlo method and the data augmentation method for the calculation of the posterior density of θ in cases where there are missing data, based on the interpretation of extensibility in the EM algorithm framework. Wei and Tanner (1990) advocated the Monte Carlo EM algorithm to perform simulations in E-step.

6 EXAMPLES

6.1 Estimation of Survival Function from Data with Censoring and Truncation: Tu et al. (1993)

It is a well-known fact that data indicating the survival time include censoring data, and many methods of analysis have been discussed for this purpose. Turnbull (1976) advocated a method of nonparametric estimation of the survival function in cases where not only censoring but also the truncating problem occurs simultaneously. Here, the EM algorithm is applied to the estimation problem regarding the survival function while tolerating the complex incompleteness associated with the general time interval data observed based on a truncated distribution on both sides, including the open intervals on both sides, in addition to the complete data in which the survival time is observed as a point of time. Tu et al. (1993) pointed out that such incompleteness has occurred in real life in the survey data of survival time relating to AIDS and estimated the survival function using the EM algorithm.

Assume that X is the calendar date on which one is diagnosed with AIDS, and S is the period between diagnosis and death. $F(s|z)$ is the cumulative distribution function of S with a given covariate z, and $f(s|z)$ is the density function. $F(s)$ and $f(s)$ represent the baseline cdf and pdf, respectively, when covariate Z is 0. The data observed in relation to individual i are (A_i, B_i, z_i). Here, A_i is the censored domain, including the survival time s_i of individual i, and B_i is the truncated domain concerning the observation of individual i. Assuming that x^* is the period between the launch of the survey and the execution of analysis, $B_i = [0, x^* - x_i]$.

Assuming that I_c is a probability index that equals 1 when $s \in C$, and equals 0 in other cases, the log likelihood of a sample of size N is:

$$\text{LL} = \sum_{i=1}^{N} \left\{ \log\left[\int_0^{s^*} I_{A_i} f(s \mid z_i) \mathrm{d}s \right] - \log\left[\int_0^{s^*} I_{B_i} f(s \mid z_i) \mathrm{d}s \right] \right\} \qquad (17)$$

In Eq. (17), the survival time observed in the sample is limited to the discrete point $s_0, s_1, \ldots, s_J(s_J = s^*)$. In this case, it is presumed that $s_j = j(0 \leq j \leq J)$ especially because the survival time is observed on a quarterly basis.

Assuming that $\xi_{ij} = 1$ when $j \in A_i$ and $\xi_{ij} = 0$ in other cases, and that $\eta_{ij} = 1$ when $j \in B_i$ and $\eta_{ij} = 0$ in other cases, Eq. (17) changes as follows:

$$LL = \sum_{i=1}^{N} \left\{ \log \left[\sum_j \xi_{ij} f(j \mid z_i) \right] - \log \left[\sum_j \eta_{ij} f(j \mid z_i) \right] \right\} \tag{18}$$

If the data are neither censored nor truncated, these become:

$$LL = \sum_{i=1}^{N} \sum_{j=1}^{J} \left\{ I_{ij} \log f(j \mid z_i) \right\} \tag{19}$$

where I_{ij} is an index that equals 1 when $s = j$ in regard to the survival time of the ith individual, and equals 0 in other cases.

Assume that the discrete proportional hazard model is:

$$f(j \mid z) = \begin{cases} (p_0 \ldots p_{j-1})^{\exp\{z^T \beta\}} \left(1 - p_j^{\exp\{z^T \beta\}} \right) & \text{if } 0 \leq j \leq J-1 \\ (p_0 \ldots p_{j-1})^{\exp\{z^T \beta\}} & \text{if } j = J \end{cases} \tag{20}$$

where $p_j = Pr(S \geq j+1) \mid S \geq j)$ $(0 \leq j \leq J-1)$ represents the conditional baseline probability corresponding to $z = 0$. If $a_j = \log[-\log(p_j)]$ $(1 \leq j \leq J-1)$, the assumed model indicates that the hazard function $h(s \mid z)$ is approximately represented by:

$$h(j \mid z) \approx \exp\{\alpha_j + z^T \beta\} \tag{21}$$

Assuming that $\theta = (\alpha, \beta)$, the maximum likelihood estimation is performed with respect to parameter θ based on the EM algorithm. J_{ij} represents the number of individuals who indicate survival time s_j among the individuals with covariate z_i.

E-step: The conditional expectation of I_{ij} for each censored individual given the previous $\theta^{(k)}$ and observed data Y_{obs} is:

$$c_{ij}^{(k)} \equiv E\left(I_{ij} \mid Y_{obs}, \theta^{(k)} \right) = \xi_{ij} \frac{f\left(j \mid z_i, \theta^{(k)} \right)}{\sum_{r=0}^{J} \xi_{ir} f\left(r \mid z_i, \theta^{(k)} \right)} \tag{22}$$

for $1 \leq i \leq N$ and $0 \leq j \leq J$. Similarly, for the truncation, we need the conditional expectation of J_{ij}. If we suppose that $\Sigma_{r=0}^{J}(1-\eta_{ij})J_{ir}$ follows a negative binomial distribution:

$$NB[m \mid n_i, P(z_i)] = \binom{m + n_i - 1}{m}[1 - P(z_i)]^m P(z_i)^{n_i} \tag{23}$$

then for $1 \leq i \leq N, 0 \leq j \leq J$, the conditional expectation of J_{ij} is:

$$g_{ij}^{(k)} = E\left(J_{ij} \mid Y_{\text{obs}}, \boldsymbol{\theta}^{(k)}\right) = \frac{(1 - \eta_{ij})f\left(j \mid z_i, \boldsymbol{\theta}^{(k)}\right)}{\sum_{r=0}^{J} \eta_{ir}f\left(r \mid z_i, \boldsymbol{\theta}^{(k)}\right)}, \tag{24}$$

where $P(z_i) = \Sigma_{r=0}^{J} \eta_{ir}f(r \mid z_j, \boldsymbol{\theta}^{(k)})$, $n_i = \Sigma_{r=0}^{J} \eta_{ir}J_{ir}$.

M-step: To update $\boldsymbol{\theta}^{(k+1)}$ by maximizing the expected complete data log likelihood:

$$LL^*\left(\boldsymbol{\theta} \mid Y_{\text{obs}}, c_{ij}^{(k)}, g_{ij}^{(k)}\right) = \sum_{i=1}^{N} \sum_{j=1}^{J}\left[c_{ij}^{(k)} + g_{ij}^{(k)}\right]\log f\left(j \mid z_i, \boldsymbol{\theta}^{(k)}\right) \tag{25}$$

Tu et al. (1993) also gave a method for the calculation of the asymptotic variance using the method of Louis (1982) or the SEM algorithm (Meng and Rubin, 1991).

6.2 Repeated-Measures Model for the Data with Missing Values

In this section, we consider a repeated-measures model for the data with missing values:

$$Y_i \sim N(X_i^* \beta, \Sigma(\boldsymbol{\theta}))(i = 1, \dots, n) \tag{26}$$

where Y_i is a $T \times 1$ random vector, X_i^* is a known $T \times p$ design matrix, and β is a $T \times 1$ parameter vector [see Table 1 in Jennrich and Schluchter (1986) for examples of covariance structures model $\Sigma(\boldsymbol{\theta})$].

Now, we write $Y_i' = (y_i^{(0)'}, Y_i^{(1)'})$, where $Y_i^{(1)}$ denotes the observed values and $Y_i^{(0)}$ denotes the missing values.

The error term and covariance matrix are partitioned corresponding to $y_i' = (y_i^{(o)'}, y_i^{(m)'})$, as follows:

$$e_i = \begin{pmatrix} e_i^{(0)} \\ e_i^{(1)} \end{pmatrix}, \Sigma = \begin{pmatrix} \Sigma_{00,i} & \Sigma_{01,i} \\ \Sigma_{10,i} & \Sigma_{11,i} \end{pmatrix}$$

The EM algorithm for this model is described as follows:

E-step:

$$E\left(e_i^{(0)} \mid Y_i^{(1)}, x_i, y_{\text{obs}}, \beta^{(t)}, \Sigma^{(t)}\right) = \Sigma_{01,i} \Sigma_{11,i}^{-1} e_i^{(1)} \equiv \widehat{e}_i^{(0)} \tag{27}$$

$$E\left(e_i^{(0)'} e_i^{(0)} \mid Y_i^{(1)}, x_i, \beta^{(t)}, \Sigma^{(t)}\right) = \widehat{e}_i^{(0)'} \widehat{e}_i^{(0)} + \Sigma_{21,i} \Sigma_{12,i} \tag{28}$$

M-step:

$$\beta^{(t+1)} = \left(\sum_{i=1}^{n} X_i' \Sigma_{11,i}^{-1} X_i\right)^{-1} \sum_{i=1}^{n} X_i' \Sigma_{11,i}^{-1} Y_i \tag{29}$$

$$\Sigma^{(t+1)} = \sum_{i=1}^{n} (\widehat{e}_i' \widehat{e}_i + R_i) \tag{30}$$

where:

$$\widehat{e} = \begin{pmatrix} e^{(1)} \\ \widehat{e}^{(0)} \end{pmatrix}, R_i = \begin{pmatrix} 0 & 0 \\ 0 & \Sigma_{22,i} - \Sigma_{22,i} \Sigma_{11,i}^{-1} \Sigma_{12,i} \end{pmatrix} \tag{31}$$

Using the GEM algorithm, θ can be estimated. Jennrich and Schluchter (1986) used a scoring method in the M-step to get $\theta^{(k+1)}$ such that the complete data log likelihood at $\theta^{(k+1)}$ is larger than that at $\theta^{(k)}$.

Yamaguchi (1990) proposed a robust model using a scale mixture of normal distributions assumption for such data. The next chapter describes such robust model.

REFERENCES

1. Amari, S. (1996). Information geometry of the EM and EM algorithms for neural networks. *Neur. Comput.* 7:13–18.
2. Boyles, R. A. (1983). On the convergence of the EM algorithm. *J. R. Stat. Soc. B* 45:47–50.
3. Dempster, A. P., Laird, N. M., Rubin, D. B. (1977). Maximum likelihood from incomplete data via the EM algorithm (with discussion). *J. R. Stat. Soc. B* 39:1–38.
4. Efron, B., Hinkley, D. V. (1978). Assessing the accuracy of the maximum likelihood estimator: observed versus expected Fisher information. *Biometrika* B65:457–487.
5. Geman, S., Geman, D. (1984). Stochastic relaxation, Gibbs distributions,

and the Bayesian restoration of images. *IEEE Trans. Pattern Anal. Mach. Intell.* 6:721–741.

6. Jamshidian, M. (1997). *t*-Distribution modeling using the available statistical software. *Comput. Stat. Data Anal.* 25:181–206.

7. Jennrich, R. I., Schluchter, M. D. (1986). Unbalanced repeated-measures models with structured covariance matrices. *Biometrics* 42:805–820.

8. Laird, N. W. (1978). Empirical Bayes methods for two-way tables. *Biometrika* 69:581–590.

9. Lange, K. (1995). A gradient algorithm locally equivalent to the EM algorithm. *J. R. Stat. Soc. B* 57:425–437.

10. Liu, C., Rubin, D. B. (1994). The ECME algorithm: a simple extension of EM and ECM with faster monotone convergence. *Biometrika* 81:633–648.

11. Liu, C., Rubin, D. B. (1995). ML estimation of the *t* distribution using EM and its extensions, ECM and ECME. *Stat. Sinica* 5:19–39.

12. Louis, T. A. (1982). Finding the observed information matrix when using the EM algorithm. *J. R. Stat. Soc. B* 44:226–233.

13. McLachlan, G. J., Krishnan, T. (1997). *The EM Algorithm and Extension.* New York: Wiley.

14. Meng, X. L. (1997). The EM algorithm and medical studies: a historical link. *Stat. Meth. Med. Res.* 6:3–23.

15. Meng, X. L., Dyk, D. (1997). The EM algorithm—an old folk song sung to a fast new tune. *J. R. Stat. Soc. B* 59:511–567.

16. Meng, X. L., Pedlow, S. (1992). EM: a bibliographic review with missing articles. *Proceedings of the Statistical Computing Section, ASA.* Alexandria, VA: ASA, pp. 24–27.

17. Meng, X. L., Rubin, D. B. (1991). Using EM to obtain asymptotic variance-covariance matrices—the SEM algorithm. *J. Am. Stat. Assoc.* 86:899–909.

18. Meng, X. L., Rubin, D. B. (1993). Maximum likelihood estimation via the ECM algorithm: a general framework. *Biometrika* 80:267–278.

19. Orchard, T., Woodbury, M. A. (1972). A missing information principle: theory and applications. *Proceedings of the 6th Berkeley Symposium on Mathematical Statistics and Probability* Vol. 1:697–715.

20. Rubin, D. B. (1987a). *Multiple Imputation for Nonresponse in Surveys.* New York: Wiley.

21. Rubin, D. B. (1987b). The SIR algorithm. *J. Am. Stat. Assoc.* 82:543–546.

22. Rubin, D. B. (1991). EM and beyond. *Psychometrika* 56:241–254.

23. Steel, B. M. (1996). A modified EM algorithm for estimation in generalized mixed models. *Biometrics* 52:1295–1310.

24. Tanner, M. A., Wong, W. H. (1987). The calculation of posterior distributions by data augmentation. *J. Am. Stat. Assoc.* 82:528–550.

25. Tu, X. M., Meng, X. L., Pagano, M. (1993). The AIDS epidemic: estimating survival after AIDS diagnosis from surveillance data. *J. Am. Stat. Assoc.* 88:26–36.

26. Turnbull, B. W. (1976). The empirical distribution with arbitrarily grouped, censored and truncated data. *J. R. Stat. Soc. B* 38:290–295.

27. Watanabe, M. (1989). Analytical methods of questionnaires for consumers' preferences—latent class analysis for various types of data. *Quality for Progress and Development*. New Delhi: Western Wiley, pp. 683–689.

28. Wei, G. C. C., Tanner, M. A. (1990). A Monte Carlo implementation of the EM algorithm and the poor man's data augmentation algorithms. *J. Am. Stat. Assoc.* 85:699–704.

29. Wu, C. F. J. (1983). On the convergence properties of the EM algorithm. *Ann. Stat.* 11:95–103.

30. Yamaguchi, K. (1990). Analysis of repeated measures data with outliers. *Bull. Inform. Cybern.* 24:71–80.

31. Yamaguchi, K. (1991). Efficiency of robust methods in factor analysis. *Proceedings of the China–Japan Symposium on Statistics*. Yunnan: Yunnan University, pp. 420–423.

32. Yamaguchi, K., Watanabe, M. (1991). A robust method in factor analysis. *Symbolic–Numeric Data Analysis and Learning*. New York: Nova Science, pp. 79–87.

4
Robust Model and the EM Algorithm

Kazunori Yamaguchi
Rikkyo University, Tokyo, Japan

1 LINEAR MODEL WITH HEAVY-TAILED ERROR DISTRIBUTIONS

Error terms in most statistical models are assumed to be the random variables following the normal distribution and, under this assumption, the maximum likelihood estimation is performed. In this case, the theoretical validity of the results is guaranteed only when data satisfy the assumption of normality. In the general case of applying such a method to actual data, its robustness becomes a problem.

As a model of error distribution other than the normal distribution, a scale mixture of normals might be used, which has a relatively heavier tail than that of the normal and is unimodal and symmetrical in distribution. The family of scale mixtures of the normal distribution includes, in particular, *t*-distribution, double exponential distribution, and logistical distribution.

The assumption of a heavier-tailed distribution reflects an interest in estimates, which are relatively unaffected by outliers. In particular, *t*-distribution has been frequently used in the analysis of real data (Zellner, 1976; Sutradhar and Ali 1986), when they considered that data included some outliers. Aitkin and Wilson (1980) treated several types of mixture models of two normals. In this section, we do not confine the *t*-family or contaminated normal family, but instead employ the family of scale mixtures of the normals and give a general method for parameter estimation.

At that time, two problems would arise. One is the identification of error distribution in the family, and the other is the detection of outliers. As mentioned in Chapter 1, we treat both problems as model selection with the help of the AIC.

1.1 Linear Model and Iteratively Reweighted Least Squares

We begin with the linear model and iteratively reweighted least squares (IRLS). The data consist of an $n \times 1$ response vector Y and an $n \times m$ design matrix X. It is assumed that:

$$Y = X\beta + e$$

where β is a vector of parameters and e is a vector such that the components s_i of $\sigma^{-1}e$ are independently and identically distributed with known density $f(s_i)$ on $-\infty < s_i < \infty$. In the context of ordinary least squares, we do not use the assumption of error distribution. The weighted least squares estimate of β is chosen to minimize:

$$(Y - X\beta)'W(Y - X\beta) \tag{1}$$

for a particular given W, where W is a positive definite diagonal matrix. We assume that $X'WX$ is full rank, so that the unique solution that attains the minimum of Eq. (1) exists and can be written as:

$$b(W) = (X'WX)^{-1}(X'WY)$$

As the weighted least squares estimate depends on the weight matrix W, we have to select a proper weight matrix. When the weight matrix is not fixed, IRLS is used. IRLS is a process of obtaining the sequence $b^{(0)}, b^{(1)}, \ldots, b^{(l+1)}$ for $l > 0$ is a weighted least squares estimate corresponding to a weight matrix $W^{(l)}$, where $W^{(l+1)}$ depends on $b^{(l)}$. To define a specific version of IRLS, we need to define a sequence of weight matrices.

A general statistical justification for IRLS arises from the fact that it can be viewed as an ML estimation.

The log likelihood is:

$$\ell(\beta, \sigma) = -n \log \sigma \sum_{i=1}^{n} \log f(s_i) \tag{2}$$

where

$$s_i = (y_i - X_i\beta)/\sigma$$

Let:

$$
w(s) - \left\{
\begin{array}{ll}
-\dfrac{df(s)/ds}{sf(s)}, & \text{for } s \neq 0 \\[3mm]
-\lim_{s \to 0} \dfrac{df(s)/ds}{sf(s)}, & \text{for } s = 0
\end{array}
\right\}
\tag{3}
$$

We assume in Eq. (3) that $f(s) > 0$ for all s, that $df(s)/ds$ exists for $z \neq 0$, and that $w(s)$ has a finite limit as $z \to 0$. In addition, because $w(s)$ is selected as the weight function, we must assume that $df/ds \leq 0$ for $s > 0$ and $df/ds \geq 0$ for $s < 0$, hence $f(s)$ is unimodal with a mode at $s = 0$. Furthermore, to simplify the theory, we assume that $df(0)/ds = 0$.

Dempster et al. (1980) gave the following lemmas and theorems concerned with the connection between the IRLS process and the log likelihood function (Eq. (2)).

Lemma 1 *For (β,σ) such that $\sigma > 0$ and $w(z_i)$ are finite for all i, the equations derived from the log likelihood (Eq. (2)) are given by:*

$$
X'WY - X'WX\beta = 0
\tag{4}
$$

and

$$
-(Y - X\beta)'W\,(Y - X\beta) + n\sigma^2 = 0
\tag{5}
$$

where W is a diagonal matrix with elements $w(s_1), w(s_2), \ldots, w(s_n)$.

Lemma 1 suggests an IRLS procedure. Because W depends on β and σ, we cannot immediately solve Eqs. (4) and (5). Thus we might derive an iterative procedure: At each iteration, substitute the temporary values of β and σ into the expression for W; then, holding W fixed, solve Eqs. (4) and (5) to obtain the next values of β and σ, that is, we take:

$$
\hat{\beta}^{(l+1)} = (X'W^{(l)}X)^{-1}X'W^{(l)}Y
\tag{6}
$$

and

$$
\hat{\sigma}^{(l+1)} = (Y - X\hat{\beta}^{(l+1)})'W^{(l)}(Y - X\hat{\beta}^{(l+1)})/n
\tag{7}
$$

Theorem 1 *If an instance of an IRLS algorithm defined by Eqs. (6) and (7) converged to (β^*,σ^*), where the weights are all finite and $\sigma^* > 0$, then (β^*,σ^*) is a stationary point of (β,σ).*

1.2 Scale Mixtures of Normal Distributions

If u is a standard normal random variable with density:

$$(2\pi)^{-1/2}\exp\left(-\frac{1}{2}u^2\right) \quad (-\infty < u < \infty)$$

and q is a positive random variable distributed independently of u with distribution function $M(q)$, then the random variable $z = uq^{-1/2}$ is called to have a scale mixture of normal distributions. Andrew et al. (1972) states it had a normal/independent distribution.

The scale mixtures of normal distributions are a convenient family of symmetrical distributions for components of error terms. The following show some familiar examples of these.

Example 1 Contaminated normal distribution

If

$$M(q) = \begin{cases} 1 - \delta & \text{if } q = 1 \\ \delta & \text{if } q = 1 \\ 0 & \text{otherwise} \end{cases}$$

then the distribution of z is the contaminated normal distribution with contaminated fraction δ and variance inflation factor λ, that is,

$$z \sim (1 - \delta) \times N(0, \Sigma) + \delta \times N(0, \Sigma/\lambda)$$

Example 2 t-distribution
Let v be a constant. If the distribution of $v \times q$ is χ^2 distribution with v degrees of freedom, then the distribution of z is the t-distribution with v degrees of freedom. When $v = 1$, it is also called the Cauchy distribution.

Example 3 Double exponential distribution

If

$$M(q) = \frac{1}{2}q^{-2}\exp\left(\frac{-1}{2q}\right)$$

z has the double exponential distribution with the probability density function:

$$f(z) = \frac{1}{2}\exp(-|z|)$$

Example 4 Logistic distribution

If

$$M(q) = \sum_{k=1}^{\infty} (-1)^{(k-1)} k^2 q^{-2} \exp\left(-\frac{k}{2q}\right)$$

z has the logistic distribution with the distribution function:

$$F(z) = [1 + \exp(-x)]^{-1}$$

Dempster et al. (1980) pointed out a close connection with IRLS. Knowledge of the scale factors $q_i^{-1/2}$ in each component $e_i = \sigma u_i q_i^{-1/2}$ would lead to the use of weighted least squares with a weight matrix W whose diagonal elements are $q_1, q_2, \ldots, q_n$, and treating these weights as missing data might lead to a statistically natural derivation of IRLS.

The density function of z, $f(z)$, is:

$$f(z) = \int_0^\infty (2\pi)^{-1/2} q^{1/2} \exp\left(-\frac{1}{2} q z^2\right) dM(z) \tag{8}$$

Lemma 2 *Suppose that z is a scale mixture random variable of normal distribution with density function (Eq. (8)). Then for $0 < |z| < \infty$:*

> *(i) The conditional distribution of q given z exists.*
> *(ii) $E(q^k|z) < \infty$, for $k > -1/2$.*
> *(iii) $w(z) = E(q|z)$.*
> *(iv) $dw(z)/dz = -z \mathrm{var}(q|z)$.*
> *(v) $w(z) = w(-z)$ is finite, positive, and nonincreasing for $z > 0$.*
> *For $z = 0$:*
> *(vi) The conditional distribution of q given z exists if and only if*
> $$E(q^{1/2}) < \infty.$$
> *(vii) $w(0) \geq w(z)$ for $z \neq 0$ and $w(0)$ is finite if and only if $E(q^{3/2})$*
> $$< \infty.$$
> *(viii) $dw(0)/dz$ is finite if and only if $E(q^{5/2}) < \infty$.*

Lemma 3 *Suppose that $u \sim N(0,1)$ and that q is a positive random variable distributed independently of u with distribution function $M(q)$. Then:*

$$z = u q^{-1/2}$$

is equivalent to the conditional distribution of z given $q = q_0$ is $N(0, 1/q_0)$.

Lemma 4 *The kurtosis of z is never less than that of u.*

Proof

$$\frac{E(z^4)}{E(z^2)^2} = \frac{E(u^4 q^{-2})}{E(u^2 q^{-1})^2}$$

$$= \frac{E(u^4)E(q^{-2})}{E(u^2)^2 E(q^{-1})^2}$$

$$\geq \frac{E(u^4)}{E(u^2)^2}$$

The family of scale mixtures of the normal is heavier-tailed than the normal distribution in the meaning of the kurtosis. We note that the condition of the normality of u is not necessary in the above lemma; that is, even when u is not limited to a normal random variable, the tail becomes heavier than that of the distribution of the original random variable.

1.3 Multivariate Model

We now consider an extension of the above results to the multivariate case.

Basic Statistics

Let U be a p-component random vector distributed as $N(0,\Sigma)$ and let q be a positive random variable distributed independently of U with distribution function $M(q)$. Then the random vector $Z = Uq^{-1/2}$ has a scale mixture distribution of multivariate normal.

Lemma 5 *The density function of Z, $f(Z)$, is*

$$f(Z) = \int_0^\infty (2\pi)^{-1/2} q^{p/2} |\Sigma|^{-1/2} \exp\left(-\frac{1}{2} q Z'\Sigma^{-1} Z\right) dM(q) \qquad (9)$$

The mean vector and covariance matrix of Z are represented by the following lemma.

Lemma 6

$$E(Z) = 0$$

$$\text{Cov}(Z) = \int_0^\infty q^{-1} dM(q) \Sigma$$

The next lemma gives the same result as the multivariate case of Lemma 4. The multivariate kurtosis $\kappa_{2,p}$ is defined by Mardia (1970) as follows.

Let X be an arbitrary p-dimensional random vector, let μ be its $p \times 1$ mean vector, and let Σ be its $p \times p$ covariance matrix. Then:

$$\kappa_{2,p} = E\left\{[(X - \mu)'\Sigma^{-1}(X - \mu)]^2\right\}$$

Lemma 7 *The multivariate kurtosis of Z is not less than that of U.*

Proof

$$E[Z'\{E(ZZ')\}^{-1}Z)^2] = E\left[\left(\frac{Z'}{q^{1/2}}\left\{E\left(\frac{1}{q}ZZ\right)\right\}^{-1}\frac{Z}{q^{1/2}}\right)^2\right]$$

$$= E\left[\left(\frac{1}{q}Z'\left\{E\left(\frac{1}{q}\right)E(ZZ')\right\}^{-1}Z\right)^2\right]$$

$$= E[Z'\{E(ZZ')\}^{-1}Z]^2 E\left(\frac{1}{q^2}\right)\left\{E\left(\frac{1}{q}\right)\right\}^{-2}$$

$$\geq E[Z'\{E(ZZ')\}^{-1}Z]^2$$

Lemma 8

$$E(q|Z) = -Z'\frac{\mathrm{d}f(Z)}{\mathrm{d}Z}/Z'\Sigma^{-1}Zf(Z)$$

and if $Z'\Sigma^{-1}Z = Z_0'\Sigma^{-1}Z_0$, then $E(q|Z) = E(q|Z_0)$.

Let $s^2 = s^2(Z) = Z'\Sigma^{-1}Z$ and $w(s^2) = E(q|Z)$ because w has the same value if s^2 is the same.

Lemma 9 *$w(s^2)$ is finite, positive, and nonincreasing for $s^2 \neq 0$.*

Proof

$$\frac{\mathrm{d}w}{\mathrm{d}s^2} = -\frac{1}{2}\frac{\int_0^\infty q^2(2\pi)^{-1/2}q^{1/2}|\Sigma|^{-1/2}\exp\left(-\frac{1}{2}qs^2\right)\mathrm{d}M(q)}{\int_0^\infty (2\pi)^{-1/2}q^{1/2}|\Sigma|^{-1/2}\exp\left(-\frac{1}{2}qs^2\right)\mathrm{d}M(q)}$$

$$+\frac{1}{2}\left\{\frac{\int_0^\infty q(2\pi)^{-1/2}q^{1/2}|\Sigma|^{-1/2}\exp\left(-\frac{1}{2}qs^2\right)\mathrm{d}M(q)}{\int_0^\infty (2\pi)^{-1/2}q^{1/2}|\Sigma|^{-1/2}\exp\left(-\frac{1}{2}qs^2\right)\mathrm{d}M(q)}\right\}^2$$

$$\leq 0$$

We now consider the multivariate regression to give the relation between the conditional expectation of q given Z and IRLS.

Suppose $Y_1, Y_2, \ldots, Y_n$ are a set of n observations, with Y_i following the model:

$$Y_i = \beta X_i + e_i \tag{10}$$

where β is a $p \times m$ matrix of parameters, X_i is a known design matrix, and e_i / σ is a vector with the density function $f(e_i)$ and $\Sigma' = \sigma^2 I_p$ for the sake of simplicity. Then the log likelihood function is:

$$\ell(\beta) = -\frac{np}{2} \log \sigma^2 + \sum_{i=1}^{n} \log f(e_i) \tag{11}$$

Let:

$$w_{ij}^* = -\frac{1}{e_{ij} f(e_i)} \frac{\mathrm{d} f(e_i)}{\mathrm{d} e_i}$$

and

$$W_i^* = \mathrm{diag}\{w_{i1}^*, w_{i2}^*, \ldots, w_{ip}^*\}$$

where e_{ij} is the jth component of e_i. Then the likelihood equations from Eq. (11) are:

$$\sum_{i=1}^{n} W_i^* e_i X_i' = 0$$

and

$$np\sigma^2 - \sum_{i=1}^{n} e_i' W_i^* e_i = 0$$

Lemma 10 *If e_i has the density function (Eq. (3)), then:*

$$E(q_i|e_i) = e_i' W_i^* e_i / e_i' e_i$$

Although w_{ij}^* is a weight of the jth component of the ith individual, $E(q_i|e_i)$ might be regarded as a weight of the ith individual, which is regarded as a weighted average of w_{ij}^*.

1.4 ML Estimation and EM Algorithm

We now establish a concrete procedure of ML estimation using the EM algorithm. It is assumed that $Y_i = \beta X_i + e_i$ $(i = 1, \ldots, n)$, and e_i is independently identically distributed from a scale mixture of multivariate normal. Namely, there exist n mutually independent positive random variables q_i, which follow the distribution function $M(q_i)$ and, conditional on q_i, e_i follows $N(0, \Sigma/q_i)$. According to the method of description of the EM algorithm by Dempster et al. (1977), $\{Y_i\}$ represents the directly observed data, called incomplete data because further potential data $\{q_i\}$ that are not observed are assumed to exist, and we denote a representation of the complete data by $\{O_i\} = \{Y_i, q_i\}$, including both observed and unobserved data. The log likelihood of $\{O_i\}$ is:

$$\ell(\beta, \Sigma) = \text{const} - \frac{n}{2}\log|\Sigma| - \frac{1}{2}\sum_{i=1}^{n} q_i (Y_i - \beta X_i)' \Sigma^{-1} (Y_i - \beta X_i) \quad (12)$$

The evaluation of the conditional expectation of $W(\beta, \Sigma)$ (Eq. (12)) is realized in E-step, and the maximization of $E(W|Y_i)$ with respect to the objective parameters is realized in M-step.

E-step: With the observed data Y_i and the temporary values of parameters $\beta^{(l)}$ and $\Sigma^{(l)}$, the conditional expectation of W is evaluated. In this case, it is none other than determining the conditional expectation of q_i, which would be determined if $M(.)$ is specified.

M-step: Assuming that $q_i^{(l+1)}$ determined in the E-step was given, the estimated values of the parameters are renewed such that these values maximize the temporary log likelihood. In this case:

$$\hat{\beta}^{(l+1)} = \sum_{i=1}^{n} q_i^{(l+1)} Y_i X_i' \left(\sum_{i=1}^{n} q_i^{(l+1)} X_i X_i' \right)^{-1}$$

$$\hat{\Sigma}^{(l+1)} = \sum_{i=1}^{n} q_i^{(l+1)} (Y_i - \hat{\beta}^{(l+1)} X_i)(Y_i - \hat{\beta}^{(l+1)} X_i)' / n$$

The above E-step and M-step are repeatedly carried out, taking the proper initial values, and the ML estimation is obtained. The E-step is materialized by specifying the distribution of q. Hereafter, several examples are shown.

Example 5 Contaminated multivariate normal distribution

$$M(q) = \begin{cases} 1 - \delta & \text{if } q = 1 \\ \delta & \text{if } q = \lambda \\ 0 & \text{otherwise} \end{cases}$$

where $\lambda < 0$ and $0 < \delta < 1$.

When $M(q)$ is specified as described above, it assumes the distribution of the mixture of $N(\mathbf{o}, \Sigma)$ and $N(\mathbf{o}, \Sigma/\lambda)$ in the ratio of $1-\delta$ to δ.

Here, the conditional distribution of q when Y, X, and the temporary values of the parameters $\beta^{(l)}$ and $\hat{\Sigma}^{(l)}$ are given is concretely evaluated, and:

$$
\begin{aligned}
w^{(l+1)} &= E(q|e) \\
&= \frac{1 - \delta + \delta\lambda^{1+p/2}\exp\{(1 - \lambda)d^2/2\}}{1 - \delta + \delta\lambda^{p/2}\exp\{(1 - \lambda)d^2/2\}}
\end{aligned}
\tag{13}
$$

is obtained, where:

$$
d^2 = (Y - \hat{\beta}^{(l)}X)'\hat{\Sigma}^{(l)-1}(Y - \hat{\beta}^{(l)}X)
$$

Example 6 Multivariate t-distribution

If $q \times v$ has a chi-square distribution with v degrees of freedom, the marginal distribution is the multivariate t-distribution (Cornish, 1954). At this time:

$$
\begin{aligned}
w^{(l+1)} &= E(q|e) \\
&= (v + p)/(v + d^2)
\end{aligned}
\tag{14}
$$

Both models downweight observations with large d^2. However, the curve of the weights is quite different for the two models—with the multivariate t-model producing relatively smoothly declining weights with increasing d^2, and the contaminated normal model tending to concentrate on the low weights in a few outlying observations.

Estimation of Mixing Parameters

If the model of the distribution function of q includes some unknown parameters (e.g., the degrees of freedom v for the multivariate t-model, the contamination fraction δ and variance inflation factor λ for the contaminated normal model, etc.), we have to estimate such parameters.

The distribution of p-variate random vector Y is assumed such that the conditional distribution of Y given positive random variable q is $N(\mu, \Sigma/q)$. Let $f(q; \theta)$ be the probability density function of q with unknown parameter vector θ (mixing parameters), and let $g(Y; \mu, \Sigma)$ denote the normal density function with mean vector μ and covariance matrix Σ. Then the joint density function of Y and q is $f(q; \theta)g(Y; \mu, \Sigma/q)$. Because θ is not included in $g(\cdot)$ but in $f(\cdot)$, the log likelihood concerned with θ, based on complete data $\{(Y_i, q_i), i = 1, 2, \ldots, n\}$, is:

$$
\text{const} + \sum_{i=1}^{n} \log f(q_i; \theta)
\tag{15}
$$

The ML estimation of θ is performed via the EM algorithm: The evaluation of the conditional expectation of Eq. (15), given observations $\{Y_i, i = 1, \ldots, n\}$ and temporary values of parameters, is realized in E-step. The maximization of the expected log likelihood obtained in E-step, with respect to θ, is realized in M-step.

We now illustrate a concrete algorithm for the t-model (Lange et al., 1989). Given lth estimates $\mu^{(l)}$, $\Sigma^{(l)}$, and $v^{(l)}$ in the E-step, we compute $w_i^{(l)}$ using Eq. (14) with $v = v^{(l)}$ and:

$$v_i = E(\log q_i | e_i)$$
$$= \psi(v^{(l)}/2 + p/2) - \log(v^{(l)}/2 + d^2/2)$$

where the digamma function (psi function) is:

$$\psi(x) = \frac{d}{dx} \log\{\Gamma(x)\}$$

In the M-step, we compute $\mu^{(l+1)}$ and $\Sigma^{(l+1)}$ and find $v^{(l+1)}$ that maximizes:

$$\ell_i(v) = \frac{nv}{2} \log(v/2) - n\log\{\Gamma(v/2)\} + \left(\frac{v}{2} - 1\right) \sum_{i=1}^{n} v_i^{(l)} - \frac{v}{2} \sum_{i=1}^{n} w_i^{(l)}$$

It is easy to find the value of v that maximizes ℓ_1 using a one dimensional-search (e.g., Newton's method).

For the t-model, another method is considered. We calculate the maximized log likelihood for a fixed v, which is:

$$\ell_2 = -\frac{1}{2} \log |\hat{\Sigma}| - \frac{1}{2}(v + p)\log(1 + d_i^2/v) - \frac{1}{2} p\log(v/2)$$

$$+\log[\Gamma\{(v + p)/2\}/\Gamma\{v/2\}]$$

We can regard the maximized log likelihood as a function of the degrees of freedom v, and select the value of v as the estimate, which attains the maximum over a grid of v values.

The case of the contaminated normal model is more complicated because the model includes two parameters (Little, 1988a,b). When the variance inflation factor λ is fixed in advance, it is easy to estimate λ simultaneously with μ and Σ by a general method described earlier; that is, we only have to add the calculation of $E\{I(q_i = \lambda)|e_i\}$ to the E-step and:

$$\delta^{(l+1)} = \frac{1}{n} \sum_{i=1}^{n} E\left\{I(q_i = \lambda) \mid e_i^{(l)}\right\}$$

to the M-step, where $I(.)$ is an index function and:

$$E\left\{I(q_i = \lambda) \mid Y_i, \mu^{(l)}, \Sigma^{(l)}\right\} \frac{\delta \lambda^{p/2} \exp\{(1 - \lambda)d^2/2\}}{1 - \delta + \delta \lambda^{p/2} \exp\{(1 - \lambda)d^2/2\}}$$

When λ is treated as a parameter, the simultaneous estimation of λ and δ with β and Σ cannot be directly derived because it is meaningless to estimate λ when q_i (or w_i) is given. Thus the estimation of λ is performed based on the log likelihood ℓ_3 from the marginal distribution of Y_i, which is:

$$\ell_3 = \text{const} - \frac{n}{2} \log | \Sigma | - \frac{1}{2} \sum_{i=1}^{n} d_i^2$$

$$+ \sum_{i=1}^{n} \log\left[1 - \delta + \delta \lambda^{p/2} \exp\{(1 - \lambda)d_i^2/2\}\right]$$

Then:

$$\delta^{(l+1)} = \frac{1}{n} \sum_{i=1}^{n} E\left\{I(q_i = \lambda^{(l)}) \mid Y_i, \mu^{(l+1)}, \Sigma^{(l+1)}\right\} \tag{16}$$

and $\lambda^{(l+1)}$ is obtained as a solution of the equation:

$$\lambda^{(l+1)} = \frac{p \sum_{i=1}^{n} E\left\{I\left(q_i = \lambda^{(l+1)}\right) \mid Y_i, \mu^{(l+1)}, \Sigma^{(l+1)}\right\}}{\sum_{i=1}^{n} d_i^{2(l+1)} E\left\{I\left(q_i = \lambda^{(l+1)}\right) \mid Y_i, \mu^{(l+1)}, \Sigma^{(l+1)}\right\}} \tag{17}$$

Note that the equation (Eq. (17)) for $\lambda^{(l+1)}$ depends on $\delta^{(l+1)}$, $\mu^{(l+1)}$, and $\Sigma^{(l+1)}$, and not on $\delta^{(l)}$, $\mu^{(l)}$, and $\Sigma^{(l)}$.

On the Convergence Property

Before demonstrating the property of the above method, we briefly rewrite an outline of the GEM algorithm. Instead of the "complete data" x, we observe the "incomplete data" $y = y(x)$. Let the density functions of x and y be $f(x;\phi)$ and $g(y;\phi)$, respectively. Furthermore, let $k(x|y;\phi) = f(x;\phi)/g(y;\phi)$ be the conditional density of x given y. Then the log likelihood can be written in the following form:

$$L(b\phi') = \log g(y; \phi') = Q(\phi' \mid \phi) - H(\phi' \mid \phi)$$

where:

$$Q(\phi' \mid \phi) = E\{\log f(x; \phi') \mid y, \phi\}$$
$$H(\phi' \mid \phi) = E\{\log k(x \mid y; \phi') \mid y, \phi\}$$

and these are assumed to exist for any (ϕ, ϕ).

In general, $Q(\phi' \mid \phi) - Q(\phi \mid \phi) \geq 0$ implies that $L(\phi') - L(\phi) \geq 0$. Therefore, for any sequence $\{\phi^{(p)}\}$ generated by GEM algorithm:

$$L(\phi^{(p+1)}) \geq L(\phi^{(p)}) \tag{18}$$

This is essential to the convergence property of GEM algorithm, and the hybrid GEM algorithm must keep this property.

Let us show that the above method can generate sequences such that $L(\phi^{(p+1)}) \geq L(\phi^{(p)})$. Let ψ be unknown parameters, except λ. Given $\psi^{(p)}$ and $\lambda^{(p)}$, from the step of the GEM algorithm for μ and Σ, and Eq. (16), we obtain $\psi^{(p+1)}$. Then clearly:

$$Q(\psi^{(p+1)}, \lambda^{(p)} \mid \psi^{(p)}, \lambda^{(p)}) \geq Q(\psi^{(p)}, \lambda^{(p)} \mid \psi^{(p)}, \lambda^{(p)})$$

which yields:

$$L(\psi^{(p+1)}, \lambda^{(p)}) \geq L(\psi^{(p)}, \lambda^{(p)})$$

$\lambda^{(p+1)}$ determined by Eq. (17) satisfies:

$$L(\psi^{(p+1)}, \lambda^{(p+1)}) \geq L(\psi^{(p+1)}, \lambda^{(p)}) \geq L(\psi^{(p)}, \lambda^{(p)})$$

and therefore Eq. (18).

2 ESTIMATION FROM DATA WITH MISSING VALUES

Little (1988a) improved the method explained in Sec. 1 so that it can be used for data sets with missing values.

Now, let y_i be the data vector. We write $y_i' = (y_i^{(o)'}, y_i^{(m)'})$, where $y_i^{(o)}$ denotes the observed values and $y_i^{(m)}$ denotes the missing values.

The mean vector and covariance matrix are partitioned corresponding to $y_i' = (y_i^{(o)'}, y_i^{(m)'})$, as follows:

$$\mu = \begin{pmatrix} \mu_i^{(o)} \\ \mu_i^{(m)} \end{pmatrix}, \Sigma = \begin{pmatrix} \Sigma_i^{(oo)} & \Sigma_i^{(om)} \\ \Sigma_i^{(mo)} & \Sigma_i^{(mm)} \end{pmatrix}$$

The log likelihood based on $y_1, \ldots, y_n$ is:

$$\ell(\beta, \Sigma) = \text{const} - \frac{n}{2}\log|\Sigma| - \frac{1}{2}\sum_{i=1}^{n} q_i(Y_i^{(o)} - \mu_i^{(o)})' \Sigma^{(oo)-1}(Y_i^{(o)} - \mu_i^{(o)})$$

$$- \frac{1}{2}\sum_{i=1}^{n} q_i(Y_i^{(m)} - \mu_i^{(m)} - \Sigma_i^{(mo)}\Sigma_i^{(oo)-1}(y_i^{(o)} - \mu_i^{(o)}))'$$

$$\tilde{\Sigma}^{(oo)-1}(Y_i^{(m)} - \mu_i^{(m)} - \Sigma_i^{(mo)}\Sigma_i^{(mm)-1}(y_i^{(o)} - \mu_i^{(o)})) \quad (19)$$

where $\tilde{\Sigma}^{(mm)} = \Sigma^{(mm)} - \Sigma^{(mo)}\Sigma^{(oo)-1}\Sigma^{(om)}$. This can be regarded as a linear combination of q, $qy^{(m)}$, and $qy^{(m)}y^{(m)\prime}$. In E-step, we only need to find the expectations of such statistics to get the expectation of the above log likelihood based on the complete data.

The E-step then consists of calculating and summing the following expected values:

$$E(q_i \mid y_i^{(o)}; \mu^{(l)}, \Sigma^{(l)}) \quad (20)$$

$$E(q_i y_i^{(m)} \mid y_i^{(o)}; \mu^{(l)}, \Sigma^{(l)}) \quad (21)$$

$$E(q_i \mid y_i^{(m)} y_i^{(m)\prime}; \mu^{(l)}, \Sigma^{(l)}) \quad (22)$$

Let $w_i = E(q_i \mid y_i^{(o)}; \mu, \Sigma)$, then:

$$E(q_i y_i^{(m)} \mid y_i^{(o)}; \mu, \Sigma) = w_i \hat{y}_i^{(m)} \quad (23)$$

$$E(q_i y_i^{(m)} y_i^{(m)\prime} \mid y_i^{(o)}; \mu, \Sigma) = w_i \hat{y}_i^{(m)} \hat{y}_i^{(m)\prime} + \tilde{\Sigma}_i^{(mm)} \quad (24)$$

where $\hat{y}_i^{(m)} = \Sigma_i^{(mo)}\Sigma_i^{(oo)-1}(y_i^{(o)} - \mu_i^{(o)})$.

It has been shown above that this method is a trial in which both outliers and missing value problems will be simultaneously solved by using scale mixture models of normal distributions and EM algorithm.

In addition, such trials have been applied not only to the estimation of such basic statistics but also to other multivariate analysis models (e.g., general linear model, Lange et al. 1989; repeated measures model, Little 1988b, Yamaguchi, 1989; factor analysis model, Yamaguchi and Watanabe, 1991; etc.).

The model of distribution of q, $M(q)$, may include parameters to be estimated (Kano et al., 1993 called them tuning parameters) (e.g., the degrees of freedom of the t-model, and mixing parameters of the contaminated normal model). Lange et al. (1989) and Yamaguchi (1990) have given methods to estimate such parameters. In Chapter 6, such methods are explained as examples of the extension of the EM algorithm.

In Secs. 3 and 4, we apply the distribution model explained here to the factor analysis model or the Tobit model.

3 ROBUST ESTIMATION IN FACTOR ANALYSIS

Factor Analysis Model

Factor analysis is a branch of multivariate analysis that is concerned with the internal relationships of a set of variables when these relationships can be taken to be linear, or approximately so. Initially, factor analysis was developed by psychometricians and, in the early days, only approximate methods of estimation were available, of which the most celebrated was the centroid or simple summation method. The principal factor and minres methods are more recent approximate methods (see Harman, 1967 and his references). Efficient estimation procedures were based on the method of maximum likelihood (Lawley and Maxwell, 1963). Difficulties of computational nature were experienced, however, and it was not until the advent of electronic computers and a new approach to the solution of the basic equations by Jöreskog (1967) that the maximum likelihood approach became a feasible proposition.

An alternative approach to calculating ML estimates was suggested by Dempster et al. (1977) and has been examined further by Rubin and Thayer (1982, 1983) and Bentler and Tanaka (1983). Its use depends on the fact that if we could observe the factor scores, we could estimate the parameters by regression methods.

Bentler and Tanaka pointed out some problems on Rubin and Thayer's example and the EM algorithm for ML factor analysis. Concerning the slowness of convergence of the EM algorithm, it does not seem to be so important in recent highly developed circumstances of computers. On the other hand, the problem of the example of Rubin and Thayer seems to be due to insufficient implement of the algorithm. In particular, it is important to determine what kind of criterion for convergence is selected. In applying the EM algorithm, we have to use strict criterion for convergence, which was also noted by Bentler and Tanaka, because a small renewal of parameters is performed by one step of the iteration when the model includes a large number of parameters.

The ordinary ML factor analysis is based on the assumption that the observations follow the multivariate normal distribution. It is well known that the analysis under the normality assumption is sensitive to outliers. In fact, in practical applications of factor analysis, we often meet cases where the normality assumption is inappropriate because the data include some extreme observations. Rubin and Thayer (1982) mentioned that, after deriving the ML method for factor analysis under the multivariate normal assumption, "the entire issue of the sensitivity of results to the assumption of multivariate normality is important for the wise application of the technique in practice."

In many fields including social and behavioral sciences, the normality assumption is sometimes unrealistic. Thus it becomes important to know how robust the normal theory is to violations of the multivariate normality assumption. Shapiro and Browne (1987) show that for elliptical distributions, the maximum likelihood and least square estimates are efficient within the class of estimates based on the sample covariance matrix, and Browne (1984) gives asymptotic distribution-free (ADF) methods. However, their methods are not necessarily efficient within the class of all possible estimates. In particular, when the kurtosis of the population distribution is large, they are significantly inefficient. This fact suggests the necessity of a consideration of more efficient estimates under the distribution with large kurtosis.

3.2 Robust Model

Suppose $Y_i = \alpha + \beta Z_i + e_i$ ($i = 1, \ldots, n$), where Y_i is an observed p-component vector, Z_i is an unobserved m-component vector of factor scores and e_i is a vector of errors (or errors plus specific factors). α is a vector of means and the $p \times m$ matrix β consists of factor loadings.

In this section, we use scale mixtures of multivariate normal distributions instead of the normality assumption, considering following two typical backgrounds. In employing such heavier-tailed symmetrical distributions as underlying distributions, we consider two practical possibilities as follows:

(1) A case where a group is not homogeneous from the beginning, and the existence of partial subjects that have abnormal capability is supposed. Namely, both Y and Z are assumed to follow scale mixtures of normal distributions.

(2) A case where the latent ability itself of a group is homogeneous, but at the point of time when manifest response Y is observed, outliers mix. Originally, because specific factors are the result of the mixture of many factors including errors, the assumption for the distribution of specific factors in applying scale mixtures of normal distributions is more realistic than normal distributions.

Concrete statistical models based on the above two cases are as follows:

Model 1 We assume that, conditional on unobserved q_i, e_i is normally distributed with mean 0 and covariance matrix Ψ/q_i and Z_i is also normally distributed with mean 0 and covariance matrix I/q_i, and that e_i and Z_i are mutually independent, where Ψ is a diagonal matrix and I is the unit matrix. q_i is a positive random variable with the probability (density) function $M(q_i)$.

Then, conditional on q_i,

$$\begin{pmatrix} Y_i \\ Z_i \end{pmatrix} \sim N\left(\begin{pmatrix} \alpha \\ 0 \end{pmatrix}, \Sigma^{(1)}/q_i\right)$$

where:

$$\Sigma^{(1)} = \begin{pmatrix} \Sigma_{YY}^{(1)} & \Sigma_{YZ}^{(1)} \\ \Sigma_{ZY}^{(1)} & \Sigma_{ZZ}^{(1)} \end{pmatrix} = \begin{pmatrix} \beta\beta' + \Psi & \beta \\ \beta' & I \end{pmatrix}$$

Model 2 Z_i is, independently of q_i, normally distributed with mean 0 and covariance matrix I. Conditional on q_i, e_i is normally distributed with mean 0 and covariance matrix Ψ/q_i. Thus, conditional on q_i,

$$\begin{pmatrix} Y_i \\ Z_i \end{pmatrix} \sim N\left(\begin{pmatrix} \alpha \\ 0 \end{pmatrix}, \Sigma_i^2\right)$$

where:

$$\Sigma_i^{(2)} = \begin{pmatrix} \Sigma_{YYi}^{(2)} & \Sigma_{YZi}^{(2)} \\ \Sigma_{ZYi}^{(2)} & \Sigma_{ZZi}^{(2)} \end{pmatrix} = \begin{pmatrix} \beta\beta' + \Psi/q_i & \beta \\ \beta & I \end{pmatrix}$$

3.3 Estimation of Parameters

In this section, assuming that the number of factors is known, we give the estimates of parameters by applying the EM algorithm, treating q and Z as missing data, that iteratively maximizes the likelihood supposing q and Z were observed. First we consider the estimation under Model 1. The following lemma enables us to easily handle the log likelihood.

Lemma 11

$$\left|\Sigma^{(1)}\right| = |\Psi|$$

$$\left|\Sigma^{(2)}\right| = q^{-p}|\Psi|$$

$$\Sigma^{(1)-1} = \begin{pmatrix} \Psi^{-1} & -\Psi^{-1}\beta \\ -\beta'\Psi^{-1} & I_m + \beta'\Psi^{-1}\beta \end{pmatrix}$$

$$\Sigma^{(2)-1} = \begin{pmatrix} q\Psi^{-1} & -q\Psi^{-1}\beta \\ -q\beta'\Psi^{-1} & I_m + q\beta'\Psi^{-1}\beta \end{pmatrix}$$

If Zs and qs are observed in addition to Y, the log likelihood is:

$$\ell = \text{const} - \frac{n}{2}\log|\Psi| - \frac{1}{2}\sum_{i=1} n\text{tr}[q_i\{\Psi^{-1}((Y_i - \alpha)(Y_i - \alpha)'$$
$$-2(Y_i - \alpha)Z_i'\beta' + \beta Z_i Z_i'\beta')\}]$$

and the sufficient statistics are

$$\sum q_i, \quad \sum q_i Y_i, \quad \sum q_i Z_i, \quad \sum q_i Y_i Y_i', \quad \sum q_i Y_i Z_i'$$

Let:

$$\begin{pmatrix} S_{YY} & S_{YZ} \\ S_{ZY} & S_{ZZ} \end{pmatrix} = \begin{pmatrix} \sum q_i Y_i Y_i'/q_0 & \sum q_i Y_i Z_i'/q_0 \\ \sum q_i Z_i Y_i'/q_0 & \sum q_i Z_i Z_i'/q_0 \end{pmatrix}$$

$$\begin{pmatrix} C_{YY} & C_{YZ} \\ C_{ZY} & C_{ZZ} \end{pmatrix} = \begin{pmatrix} S_{YY} - \overline{Y}\overline{Y}' & S_{YZ} - \overline{Y}\overline{Z}' \\ S_{ZY} - \overline{Z}\overline{Y}' & S_{ZZ} - \overline{Z}\overline{Z}' \end{pmatrix}$$

where:

$$\overline{Y} = \sum q_i Y/q_0, \quad \overline{Z} = \sum q_i Z_i/q_0, \quad q_0 = \sum q_i$$

then:

$$\hat{\alpha} = \overline{Y} - \hat{\beta}\overline{Z} \tag{25}$$
$$\hat{\beta} = C_{YZ}C_{ZZ}^{-1}$$
$$\hat{\Psi} = \text{diag}(C_{YY} - C_{YZ}C_{ZZ}^{-1}C_{ZY})q_0/n$$

However, we cannot observe qs and Zs. Thus we must calculate the conditional expectations of the above sufficient statistics given Ys.

E-step: We give the conditional expectations of the sufficient statistics given Ys as follows. First we let:

$$w_i = E(q_i \mid Y_i)$$

where the specific form of w_i depends on the model for the distribution of q_i (i.e., $M(.)$; see examples).

We note that, in this case, we could regard w_i as the weight of Y_i in the following procedure. Therefore, we could easily find the extreme observations by checking w_i (see the results of Mardia et al., 1979 in Sec. 5.4):

$$E(q_i Z_i \mid Y_i) = E\{q_i E(Z_i \mid q_i, Y_i) \mid Y_i\}$$

$$= w_i \hat{Z}_i$$

because the conditional expectation of Z given q and Y does not depend on q:

$$E(q_i Z_i Z_i' \mid Y_i) = E\{q_i E(Z_i Z_i' \mid q_i, Y_i) \mid Y_i\}$$
$$= E\left\{ q_i \left(\hat{Z}_i \hat{Z}_i' + \text{Cov}(Z_i \mid q_i, Y_i) \right) \mid Y_i \right\}$$
$$= w_i \hat{Z}_i \hat{Z}_i' + \Sigma_{ZZ}^*$$

where:

$$\hat{Z}_i = \Sigma_{ZY}^{(1)} \Sigma_{YY}^{(1)-1} (Y_i - \alpha)$$
$$\Sigma_{ZZ}^* = \Sigma_{ZZ}^{(1)} - \Sigma_{ZY}^{(1)} \Sigma_{YY}^{(1)-1} \Sigma_{YZ}^{(1)}$$

M-step: We compute the update estimates with equations (Eq. (25)) replaced by their conditional expectations from E-step.

We would get the ML estimates by applying repeatedly the E-step and the M-step until convergence.

Example 6 To consider the contaminated multivariate normal case, let:

$$M(q_i) = \begin{cases} 1 - \delta & \text{if} \quad q_i = 1 \\ \delta & \text{if} \quad q_i = 1 \\ 0 & \text{otherwise} \end{cases}$$

then:

$$w_i = \frac{1 - \delta + \delta \lambda^{1+p/2} \exp\{(1 - \lambda) d_i^2 / 2\}}{1 - \delta + \delta \lambda^{p/2} \exp\{(1 - \lambda) d_i^2 / 2\}}$$

where:

$$d_1^2 = (Y_i - \alpha)' \Sigma_{YY}^{(1)-1} (Y_i - \alpha)$$

Example 7

If $q_i \times \nu$ has a the chi-square distribution with ν degrees of freedom, the marginal distribution is the multivariate t-distribution and:

$$w_i = \frac{\nu + p}{\nu + d_i^2}$$

(for examples, see Andrews et al., 1972; Andrews and Mallows, 1974). We note that w_i is decreasing with d_i in the above two examples. For

Model 2, we also get the estimates in the same way. But the calculation of the conditional expectations in E-step is more complicated because $E(Z_i|q_i, Y_i)$ depends on q_i in this case. In the following example, we show the expectations that would be needed in E-step, in the contaminated multivariate normal case.

Example 8. $M(q_i)$ is the same as that of Example 1; that is, this example is the contaminated multivariate normal case:

$$E(q_i \mid Y_i) = \frac{(1-\delta)|\Sigma|^{-1/2} + \delta\lambda|\Sigma^*|^{-1/2}D^2}{(1-\delta)|\Sigma|^{-1/2} + \delta|\Sigma^*|^{-1/2}D_i^2} = w$$

$$E(q_i Z_i \mid Y_i) = \beta'(h_{1i}\Sigma^{-1} + \lambda h_{\lambda i}\Sigma^{*-1})(Y_i - \alpha)$$

$$E(q_i Z_i Z_i' \mid Y_i) = w_i I_p + \beta'[h_{1i}\Sigma^{-1}\{I_p + (Y_i - \alpha)(Y_i - \alpha)'\Sigma^{-1}\} \\ + \lambda h_{\lambda i}\Sigma^{*-1}\{I_p + (Y_i - \alpha)(Y_i - \alpha)'\Sigma^{*-1}\}]\beta$$

where:

$$h_{1i} = \frac{(1-\delta)\mid\Sigma\mid^{-1/2}}{(1-\delta)\mid\Sigma\mid^{-1/2} + \delta\mid\Sigma^*\mid^{-1/2}} D_i^2$$

$$h_{\lambda i} = 1 - h_{1i} = \frac{\delta\mid\Sigma^*\mid^{-1/2}}{(1-\delta)\mid\Sigma\mid^{-1/2} + \delta\mid\Sigma^*\mid^{-1/2}} D_i^2$$

$$D_i^2 = \exp\left\{\frac{1}{2}(Y_i - \alpha)'(\Sigma^{-1} - \Sigma^{*-1})(Y_i - \alpha)\right\}$$

$$\Sigma = \beta'\beta + \Psi$$

$$\Sigma = \beta'\beta + \Psi/\lambda$$

3.4 Simulation Study

Our numerical model is based on a two-factor model for the open/closed book data in Mardia et al. (1979). Here, we set the order p of response data as 5 and the number m of common factors as 2 and made the following settings for the factor loading matrix and the specific variance matrix:

$$\alpha = 0$$

$$\beta' = \begin{bmatrix} 0.63 & 0.70 & 0.89 & 0.78 & 0.73 \\ 0.37 & 0.31 & 0.05 & -0.20 & -0.20 \end{bmatrix}$$

$$\Psi = \text{diag}(0.46, 0.42, 0.20, 0.35, 0.43)$$

This numerical model is based on the maximum likelihood estimates made by Mardia et al. During the generation of artificial data, we used the following four distributions for the factor scores Z_i and the error terms e_i:

(1) Multivariate normal distribution (MN)
(2) Multivariate t-distribution with 10 df (T10)
(3) Multivariate t-distribution with 4 df (T4)
(4) Contaminated multivariate normal distribution (CN):

$$0.9 \times N\left(\begin{pmatrix} \alpha \\ 0 \end{pmatrix}, \Sigma\right) + 0.1 \times N\left(\begin{pmatrix} \alpha \\ 0 \end{pmatrix}, \Sigma/0.0767\right)$$

For the four sets of artificial data based on different assumptions of underlying distribution generated from the above models, we calculated the following four MLEs:

(a) MLE on the assumption of multivariate normal distribution (normal MLE)
(b) MLE on the assumption of multivariate t-distribution with 10 df (T10-type MLE)
(c) MLE on the assumption of multivariate t-distribution with 4 df (T4-type MLE)
(d) MLE on the assumption of contaminated multivariate normal distribution (contaminated MLE).

Thus, one experiment is enough to calculate a total of 16 estimates with regard to factor loadings β and specific variances Ψ. For each of these estimates, we calculated the following estimation criteria.

The square root of the root mean squared error (RMSE) with regard to factor loadings is:

$$\left\{ \sum_{i=1}^{p} \sum_{j=1}^{m} (\hat{\beta}_{ij} - \beta_{ij})^2 / pm \right\}^{1/2}$$

where $\hat{\beta}$ is rotated to satisfy that $\hat{\beta}' \hat{\Psi} \hat{\beta}$ is diagonal.

The square root of the root mean squared error with regard to specific variances is:

$$\left\{ \sum_{i=1}^{p} \left(\hat{\Psi}_i - \psi_i \right)^2 / p \right\}^{1/2}$$

We made a simulation with different sample sizes of 200 and 400. The simulation size was 1000.

Table 1 shows a summary of RMSEs related to multivariate normal distribution, contaminated-type multivariate normal distribution, and multivariate t-distribution. These four distribution patterns are arranged in the order of small to large kurtosis. Table 2 shows multivariate kurtosis (Mardia, 1970) of these four distribution patterns, and these can be arranged in order of multivariate kurtosis as follows:

MN < T10 < CN < T4

In the rows in Table 1, we provided distribution forms of random values used in the generation of artificial data. In the columns, we provided distribution forms assumed for the maximum likelihood method. Therefore, the diagonal cell in the table shows the efficiency of each maximum likelihood method under the right assumption of underlying distribution patterns. The nondiagonal cell, on the other hand, shows the efficiency of each maximum likelihood method under wrong assumptions of underlying distributions. The figures in parentheses show the relative ratios for each row on the basis that the figure for the diagonal cell is 100. The figure in parentheses for the final row (mean) indicates the averages of all these relative rations from all rows. The smaller the relative ratio is, the more robust a specific distribution pattern is with regard to erroneous regulations.

Table 1 RMSE of Estimates of Unique Variances

	Assumed distribution			
Data	MN	T10	CN	T4
	$n = 200$			
MN	72 (100)	74 (103)	72 (101)	78 (108)
T10	79 (107)	74 (100)	81 (109)	76 (103)
CN	115 (153)	75 (101)	75 (100)	76 (101)
T4	108 (142)	78 (103)	87 (114)	76 (100)
Mean	(126)	(102)	(106)	(103)
	$n = 400$			
MN	52 (100)	55 (107)	54 (102)	57 (109)
T10	62 (106)	58 (100)	60 (104)	58 (100)
CN	93 (176)	57 (109)	52 (100)	54 (103)
T4	79 (150)	57 (109)	64 (122)	53 (100)
Mean	(133)	(106)	(107)	(103)

Table 2 Multivariate Kurtosis

Distribution	MN	T10	CN	T4
Multivariate kurtosis	99	128	383	∞

First we discuss the RMSEs of estimates of factor loadings in a comparison of multivariate normal distribution with contaminated multivariate normal distribution in Table 1. With regard to artificial data that follow the multivariate normal model, the efficiency of contaminated-type MLE almost corresponds to that of normal-type MLE. On the other hand, with regard to artificial data that follow the contaminated-type multivariate normal model, the efficiency of the same contaminated-type MLE far exceeds that of normal-type MLE. That is, when underlying distribution shifts from normal-type to contaminated-type distributions, normal MLE loses its efficiency. Contaminated-type MLE, on the other hand, proves robust with regard to the distribution slippage. This tendency is similar in estimating specific variances. But an increasing difference in robustness between the two MLEs in response to a rise in sample size is even larger than that when estimating factor loadings.

Next we compare RMSEs of estimates of factor loadings with regard to multivariate normal distribution and multivariate t-distribution. With regard to artificial data that follow the multivariate normal distribution model, we can say that the efficiency of normal MLE differs little from that of multivariate t-type MLE with 4 and 10 df. But when the sample size is as large as 200, the efficiency of multivariate t-type MLE with 4 df is slightly lower than that of the other two multivariate t-distribution model with 10 df. The efficiency of the multivariate t-type MLE with 10 and 4 df is fairly high, but the efficiency of normal MLE is comparatively lower. With regard to artificial data that follow the multivariate t-distribution model with 4 df, the efficiency of the multivariate t-type MLE with 10 df is lower, but the efficiency of normal MLE is even lower. On average, MLE robustness with regard to the slippage of assumptions of underlying distribution is the highest in the case of multivariate t-type MLE with 4 df, followed by multivariate t-type MLE with 10 df, and then by multivariate normal MLE. Normal MLE is thus least robust. That is, the robustness of the resulting maximum likelihood estimator increases in direct proportion to the multivariate kurtosis of the distribution pattern assumed in the creation of the maximum likelihood method. This tendency increases as the sample size rises. The same tendency is seen in the estimation of specific variances. The increase in difference of MLE robustness according to an increase in sample size is larger than that in the estimation of factor loadings.

The result of a comparison of the contaminated multivariate normal distribution with multivariate t-distributions is different from the cases including the multivariate normal distribution. As in the earlier case, the RMSE value corresponding to the upper triangular cell is smaller than the RMSE of the lower triangular cell, with the diagonal cell as the borderline. That is, we can say that the maximum likelihood method assuming a longer tail compared with generated data distribution affects the robustness stemming from erroneous regulations of underlying distribution in the maximum likelihood method less than the maximum likelihood method, assuming that there is a distribution pattern with a shorter tail. But, unlike the earlier example including multivariate normal distribution, the gap between contaminated-type distribution and multivariate t-distribution is not so wide.

Generally speaking, MLE under normal distribution is less robust than MLE that assumes a heavier-tailed distribution.

4 ROBUST TOBIT MODEL

4.1 Introduction

In the field of economics and sociology, we often deal with the non-negative data such as amount of money, periods of time, number of persons, ratios (e.g., household expenditure on durable goods), household income, duration of unemployment or welfare receipt, number of extra-marital affairs, ratio of unemployed hours to employed hours, etc. (see Ashenfelter and Ham, 1979).

Tobin (1958) proposed a regression model having such non-negative dependent variables, where he made several observations showing zero as the censored data at zero. By the Tobit model, which was proposed by Tobin (1958), we might get a good fit of a linear regression line to the data even if the data have a nonlinear relationship. Amemiya (1984) gave a comprehensive survey of the Tobit model and, concerning non-normality case, Arabmazar and Schmidt (1982) investigated the consequences of the misspecification of normality in the Tobit model and stated that the asymptotic bias of the ML estimators assuming the normal distribution error can be substantial. Thus, it would be of great importance to develop a robust estimate for the Tobit model.

The Tobit model is defined as follows:

$$y_i^* = x_i'\beta + e_i (i = 1, \ldots, N) \tag{26}$$

$$y_i = \begin{cases} y_i^* & \text{if } y_i^* > 0 \\ 0 & \text{if } y_i^* \leq 0 \end{cases} \tag{27}$$

where e_i follows $N(0,\sigma^2)$. It is assumed that y_i and x_i are observed, but y_i^* cannot be observed if it is negative.

To make a robust model, we use a scale mixture of normal distributions as a distribution assumption of e_i:

$$e_i \mid q_i \sim N(0, \sigma^2/q_i)$$

where q_i is a positive random variable that has a known density function.

4.2 ML Estimation by EM Algorithm

In this section, we conduct the ML estimation method for the regression coefficient β and error variance σ^2 assuming that the distribution of e_i is the t-distribution with known degrees of freedom.

We regard $\{y_i, x_i, q_i \ (i = 1, 2, \ldots, N)\}$ as the unobservable complete data. The E-step and M-step are as follows:

E-step: The log likelihood for the above complete data is:

$$\ell = \text{const} - \frac{1}{2} \sum_{i=1}^{N} (y_i - x_i\beta)^2 q_i/\sigma^2) \tag{28}$$

The E-step calculates its conditional expectation given the observed data and current values of parameters. In this case, we have to get the conditional expectations of Σq_i, $\Sigma q_i y_i^*$, and $\Sigma q_i y_i^{*2}$.

M-step: The M-step computes the updated values of β and σ^2, which are obtained as values to maximize the expectation of the complete data log likelihood ( ). Then we get:

$$\hat{\beta} = (X'WX)^{-1} X'E(Wy^*)$$

$$\hat{\sigma}^2 = \sum_{i=1}^{N} \left\{ E(q_i y_i^{*2}) - 2E(q_i y_i^*) x_i' \hat{\beta} + E(q_i)(x_i' \hat{\beta})^2 \right\} / N$$

where $W = \text{diag}(q_1, q_2, \ldots, q_N)$.

Computations in the E-Step

Computations of expectations in the E-step are very complicated. The Monte Carlo method is recommended. Wei and Tanner (1990) proposed a Monte Carlo E-step. An EM algorithm where the E-step is executed by Monte Carlo is known as a Monte Carlo EM (MCEM) algorithm.

Let Y be observed data, let Z be missing data, and let $p(.)$ be the probability density function. Here, we have to compute:

$$Q(\theta, \theta^{(k)}) = \int \log p(Z, Y \mid \theta) p(Z \mid Y, \theta^{(i)}) \mathrm{d}Z$$

in the E-step. Wei and Tanner (1990) used the following the Monte Carlo E-step for this computation:

(1) Draw $z_1, \ldots, z_m$ from $p(Z/Y, \theta^i)$.
(2) $\hat{Q}_{i+1}(\theta, \theta_i) = \Sigma_{j=1}^{m} \log p(z_j, Y/\theta)/m$.

For this algorithm, the monotonicity property is lost, but McLachlan and Krishnan (1997) stated that the algorithm with the Monte Carlo E-step could get close to a maximizer with a high probability (see also Chan and Ledolter, 1995). McCulloch (1994, 1997) gave other applications, and Booth and Hobert (1999) discussed stopping rules of the MCEM algorithm.

REFERENCES

1. Aitkin, M., Wilson, G. T. (1980). Mixture models, outliers, and the EM algorithm. *Technometrics* 22:325–331.
2. Amemiya, T. (1984). Tobit models; a survey. *J. Econom.* 24:3–61.
3. Andrews, D. F., Mallows, C. L. (1974). Scale mixtures of normal distributions. *J. R. Stat. Soc., Ser. B* 36:99–102.
4. Andrews, D. F., Bickel, P. J., Hampel, F. R., Huber, P. J., Rogers, W. M., Tukey, J. W. (1972). *Robust Estimates of Location; Survey and Advances.* Princeton: Princeton University Press.
5. Arabmazar, A., Schmidt, P. (1982). An investigation of the robustness of the Tobit estimator to non-normality. *Econometrica* 50:1055–1063.
6. Ashenfelter, O., Ham, J. (1979). Education, unemployment, and earnings. *J. Polit. Econ.* 87:99–116.
7. Bentler, P. M., Tanaka, J. S. (1983). Problems with EM algorithm for ML factor analysis. *Psychometrika* 48:247–251.

8. Booth, J. G., Hobert, J. P. (1999). Maximizing generalized linear mixed model likelihoods with an automated Monte Carlo EM algorithm. *J. R. Stat. Soc., Ser. B* 61:265–285.

9. Browne, M. W. (1984). Asymptotic distribution-free methods for the analysis of covariance structures. *Br. J. Math. Stat. Psychol.* 37:62–83.

10. Chan, K. S., Ledolter, J. (1995). Monte Carlo EM estimation for time series models involving counts. *J. Am. Stat. Assoc.* 90:242–252.

11. Dempster, A. P., Laird, N. M., Rubin, D. B. (1977). Maximum likelihood from incomplete data via the EM algorithm (with discussion). *J. R. Stat. Soc., Ser. B* 39:1–38.

12. Dempster, A. P., Laird, N. M., Rubin, D. B. (1980). Iterative reweighted least squares for linear regression when errors are normal/independent distributed. In: Krishnaiah, P. R. ed. *Multivariate Analysis V.* New York: Academic Press, pp. 35–57.

13. Harman, H. H. (1967). *Modern Factor Analysis.* Chicago: University of Chicago Press.

14. Joreskog, K. G. (1967). Some contributions to maximum-likelihood factor analysis. *Psychometrika* 32:443–482.

15. Kano, Y., Berkane, M., Bentler, P. M. (1993). Statistical inference based on pseudo-maximum likelihood estimators in elliptical populations. *J. Am. Stat. Assoc.* 88:135–143.

16. Lange, K. L., Little, R. J. A., Taylor, J. M. G. (1989). Robust statistical modeling using the *t*-distribution. *J. Am. Stat. Assoc.* 84:881–896.

17. Lawley, D. N., Maxwell, A. E. (1963). *Factor Analysis as a Statistical Method.* London: Butterworth.

18. Little, R. J. A. (1988a). Robust estimation of the mean and covariance matrix from data with missing values. *Appl. Stat.* 37:23–38.

19. Little, R. J. A. (1988b). Models for continuous repeated measures data. *Stat. Med.* 7:347–355.

20. Mardia, K. V. (1970). Measures of multivariate skewness and kurtosis with applications. *Biometrika* 57:519–530.

21. Mardia, K. V., Kent, J. T., Bibby, J. M. (1979). *Multivariate Analysis.* New York: Academic Press.

22. McCulloch, C. E. (1994). Maximum likelihood variance components estimation for binary data. *J. Am. Stat. Assoc.* 89:330–335.

23. McCulloch, C. E. (1997). Maximum likelihood algorithms for generalized linear mixed models. *J. Am. Stat. Assoc.* 92:162–170.

24. McLachlan, G. J., Krishnan, T. (1997). *The EM Algorithm and Extensions.* New York: Wiley.

25. Rubin, D. B., Thayer, D. T. (1982). EM algorithms for ML factor analysis. *Psychometrika* 47:69–76.

26. Rubin, D. B., Thayer, D. T. (1983). More on EM algorithms for factor analysis. *Psychometrika* 48:253–257.

27. Shapiro, A., Browne, M. W. (1987). Analysis of covariance structures under elliptical distributions. *J. Am. Stat. Assoc.* 82:1092–1097.

28. Sutradhar, B. C., Ali, M. M. (1986). Estimation of the parameters of a regression model with a multivariate t error variable. *Commun. Stat. Theor. Meth.* 15:429–450.

29. Tobin, J. (1958). Estimation of relationships for limited dependent variables. *Econometrica* 26:24–36.

30. Yamaguchi, K. (1989). Analysis of repeated-measures data with outliers. *Bull. Inform. Cybern.* 24:71–80.

31. Yamaguchi, K. (1990). Generalized EM algorithm for models with contaminated normal error terms. In: Niki, N. ed. *Statistical Methods And Data Analysis*. Tokyo: Scientist, Inc, pp. 107–114.

32. Yamaguchi, K., Watanabe, M. (1991). A robust method in factor analysis. In: Diday, E. Lechevallier, Y. eds. *Symbolic–Numeric Data Analysis and Learning*. New York: Nova Science, pp. 79–87.

33. Wei, G. C. G., Tanner, M. A. (1990). A Monte Carlo implementation of the EM algorithm and the poor man's data augmentation algorithms. *J. Am. Stat. Assoc.* 85:829–839.

34. Zellner, A. (1976). Bayesian and non-Bayesian analysis of the regression model with multivariate student-t error terms. *J. Am. Stat. Assoc.* 71:400–405.

5
Latent Structure Model and the EM Algorithm

Michiko Watanabe
Toyo University, Tokyo, Japan

1 INTRODUCTION

One way of interpreting the relationship measured between variates in multidimensional data is to use a model assuming a conceptual, hypothetical variate (latent factor) of a lower dimension mutually related to the observed variates. Such models are collectively referred to as latent structure models and have their origin in sociologist Lazarsfeld (7). The latent class model in sociology, the factor analysis model in psychology, and the latent trait model in education and ability measurement fields are particularly common; they have been subjected to many studies historically and are high in demand in the field of practice.

The objective of this chapter is to uniformly treat the parameter estimation methods that have been studied so far on an individual basis with respect to each latent structure model, in light of the EM algorithm by Dempster et al. (2), and demonstrate the usefulness of the EM algorithm in latent structure models based on its application to latent class, latent distance, and latent trait models in concrete terms.

2 LATENT STRUCTURE MODEL

Let $x = (x_1, x_2, \ldots, x_p)$ be an observable random vector and $y = (y_1, y_2, \ldots, y_q)$ be a latent random vector. We consider the following model:

$$f(x; \gamma) = \int_{\Omega(y)} \pi(x|y; \alpha) \phi(y|\beta) dy, \tag{1}$$

where $f(x; \gamma)$ and $\phi(y; \beta)$ are the probability density functions of manifest variable x and latent variable y, $\pi(x|y; \alpha)$ is a conditional density function of x given y, and α, β, and γ are the parameters for these density functions. $\Omega(y)$ is a sample space of y.

We usually use the assumption of local independence for $\pi(x|y; \alpha)$ as follows

$$\pi(x|y; \alpha) = \prod_{j=1}^{p} \pi_j(x_j|y; \alpha), \tag{2}$$

where π_j is a density function of latent variable x_j.

In the latent structure analysis, we assume π_j as a proper form and estimate the latent parameters α and β from the observed data x. The typical latent structure models are summarized in Table 1.

Conventionally, latent parameter estimation methods have been studied on an individual basis in regard to latent structure models. Most of them were directed at algebraically solving the relational equation (explanatory equation) between manifest parameter γ and latent parameters α and β determined based on formula (1), and at applying a Newton–Raphson-type or Flettcher–Powell-type method of iteration to the direct likelihood of data relating to the manifest variate. However, many prob-

Table 1 Latent Structure Analysis

Model	Manifest variable x	Latent variable y
Latent class model	p categorical random variables	Multinominal random variable
Latent distance model	p ordinal scale variables	Multinominal ($p+1$) random variable
Latent trait model	p ordinal scale variables	Continuous random variable
Latent profile model	p continuous random variables	

lems were left unsolved in these cases, such as the indefiniteness of the solution and the generation of improper solutions. On the other hand, the estimation of latent parameters can generally be reduced to a familiar estimation problem normally employed if the responses to the latent variates of the subject group have been observed. Also, in regard to improper solutions being generated, the problem can be avoided by estimation based on data assumed by the structure of the model itself. Based on this viewpoint, the estimation problem in a latent structure model may be substituted with a parameter estimation problem based on incomplete data, and the EM algorithm may be applied on the grounds that the data relating to the latent variates are missing. In the EM algorithm, the missing value is substituted with its provisional estimate, and the estimate of the target parameter is repeatedly updated. Accordingly, the aforementioned merit can be expected to come into full play. The following sections provide an overview of the EM algorithm and build a method of estimation in concrete terms for a latent structure model.

2.1 Problems of EM Algorithm

The EM algorithm has two weaknesses: convergence is slow compared to other iteration methods that used second partial derivatives, such as the Newton–Raphson method, and the lack of evaluation of the asymptotic variance of the estimate. On the other hand, its strengths include stability until convergence is reached, avoidance of inverse-matrix operations relating to second partial derivative matrices in each iteration stage, easy programming due to the reduction of M-step into familiar, conventional methods, the wide scope of applicable models, etc.

Convergence of the EM algorithm is regarded slow in terms of the number of iterations required for convergence. Therefore if one refers to the total calculation time as the yardstick, it is impossible to categorically claim that convergence of the EM algorithm is faster/slower than the Newton–Raphson method, etc., in practice, considering the high-order inverse matrix calculations avoided per iteration. Furthermore, the latter weakness is not necessarily inherent in the EM algorithm when the combined use of the aforementioned results is taken into account.

In many cases, the likelihood of incomplete data subject to the EM algorithm is generally complex in shape, and it is difficult to learn about the traits of its shape even if it is locally confined as to whether it is unimodal or multimodal and whether it is asymmetric or symmetric. The EM algorithm guarantees a single uniform, nondecreasing likelihood trail, from the initial value to the convergence value. By taking advantage of this characteristic, it is possible to learn something about the shape by trying as many initial values broadly distributed as possible. The EM algorithm's beauty lies in, more than anything else, its ability to develop maximum likelihood estimation methods in concrete terms relatively easily in regard to a wide range of models. In particular, it can flexibly deal with constraints among parameters. This method is especially useful in cases where other methods of estimation by substitution are not final and conclusive, such as the aforementioned latent structure model, and in cases where some kind of constraint is required among parameters to assure the identifiability of the model. A notable example of this is the application of the EM algorithm to a latent structure model in concrete terms.

3 THE EM ALGORITHM FOR PARAMETER ESTIMATION

This section applies the aforementioned EM algorithm by hypothetically treating data, in which conformity is sought in the latent structure model, as incomplete data in which the portion corresponding to the latent variate that should have been observed is missing, and builds a method of estimating the maximum likelihood of the parameter with respect to a number of typical models. In particular, the existence of the missing value will not be limited to the latent variate; its existence will also be tolerated in portions relating to the manifest variates and will be examined accordingly. Although this extension makes the method of estimation somewhat more complicated, it has the advantage of making the usable data more practicable.

3.1 Latent Class Model

Latent class analysis is a technique of analysis for the purpose of determining T heterogeneous latent classes composing a population by using the response data to p items in a questionnaire. Concretely, the size of the respective latent classes in a population and the response probability in the classes for each of the p items are estimated. This method of analysis was proposed for the first time by Lazarsfeld (7), and thereafter, many researches have been carried out in connection with the technique of estimating the latent probabilities and its evaluation. For instance, from the viewpoint of algebraically solving the accounting equation system between the latent probabilities and the manifest probabilities, Anderson (1), Green (6), Gibson (4), and others have proposed their own methods of estimation. However, it has been in the application of these methods of estimation to actual data where the occurrence of improper solutions becomes a problem. Regarding the ML method, also the iterative method based on the Newton–Raphson technique was given by Lazarsfeld and Henry (8), but the occurrence of improper solution is unavoidable. As a countermeasure to the improper solution problem in the ML method, Formann (3) proposed the method of transforming the latent probabilities. Moreover, Goodman (5) gave from the other viewpoint an iterative estimation method which derives the MLEs in the range of the proper solution and showed that under the proper condition, the estimates have the asymptotic consistency, asymptotic efficiency, and asymptotic normality.

It is assumed that there are p multichotomous variables, $x_1, x_2, \ldots, x_p$, which have C_i ($i = 1, 2, \ldots, p$) categories. Besides, other than these multichotomous variables, the existence of a multichotomous variable y having T categories is assumed. Respective individuals in a sample show the response values on the above all variables, but only the response values on the first p variables can be observed actually and the response values on y cannot be observed in reality. In this sense, y is called a latent variable, and the p variables, on which the response values are observed, are called the manifest variables.

In most cases, the items of a questionnaire, reaction tests, and so on correspond to the manifest variables, and the latent concept which explains the association among the manifest variables corresponds to the latent variable and respective classes which are regarded as heterogeneous with some scale regarding that latent concept from the T categories of the latent variable.

The relation among the variable is expressed as the following equation:

$$P(x|y) = \prod_{i=1}^{p} P(x_i|y = t),$$ (3)

which is called as "local independency."

The EM algorithm for computing estimates of $P(y = t)$ (Class size) and $P(x_i|y = t)$ (Response probability of each class) is as follows,

- E-step: to compute latent frequencies $n^*(x, y)$,

$$\hat{n}^*(x|y) = E[n^*(x, y)|n(x); \{P(y)\}, \{P(x_i|y)\}]$$

$$= n(x) \frac{P(x, y)}{P(x)},$$ (4)

where

$$P(x, y) = P(y) \prod_{i=1}^{p} P(x_i|y),$$

$$P(x) = \sum_{y} P(y) \prod_{i=1}^{p} P(x_i|y).$$

- M-step: to update $P(y = t)$ and $P(x_i|y = t)$ based on $\hat{n}^*(x,y)$;

$$\hat{P}(y) = \sum_{x} \hat{n}^*(x, y)/n, \tag{5}$$

$$\hat{P}(x_i|y) = \frac{\displaystyle\sum_{x_1,\ldots,x_{i-1},x_{i+1},\ldots,x_p} \hat{n}^*(x, y)}{\displaystyle\sum_{x} \hat{n}^*(x, y)}. \tag{6}$$

REFERENCES

1. Anderson, T. W. (1954). On estimation of parameters in latent structure analysis. *Psychometrika* 19:1–10.
2. Dempster, A. P., Laird, N. M., Rubin, D. B. (1977). Maximum likelihood from incomplete data via the EM algorithm (with Discussion). *J. R. Stat. Soc. Ser. B.* 39:1–38.
3. Formann, A. K. (1978). A note on parameter estimates for Lazarsfeld's latent class analysis. *Psychometrika* 43:123–126.
4. Gibson, W. A. (1955). An extension of Anderson's solution for the latent structure equations. *Psychometrika* 20:69–73.
5. Goodman, L. A. (1979). On the estimates of parameters in latent structure analysis. *Psychometrika* 44:123–128.
6. Green, B. F. Jr. (1951). A general solution for the latent class model of latent structure analysis. *Psychometrika* 16:151–161.
7. Lazarsfeld, P. F. (1950). The logical and mathematical foundation of latent structure analysis. Measurement and Prediction. Princeton: Princeton University Press.
8. Lazarsfeld, P. F., Henry, N. W. (1968). *Latent Structure Analysis.* Boston: Houghton Mifflin.

6

Extensions of the EM Algorithm

Zhi Geng
Peking University, Beijing, China

1 PARTIAL IMPUTATION EM ALGORITHM

In this section, we present the partial imputation EM (PIEM) algorithm which imputes missing data as little as possible. The ordinary EM algorithm needs to impute all missing data for the corresponding sufficient statistics. At the E step, the PIEM algorithm only imputes a part of missing data. Thus the PIEM algorithm not only reduces calculation for unnecessary imputation, but also promotes the convergence.

Let $\{X_1, X_2, \ldots, X_K\}$ denote the set of K random variables. Let $T = \{t_1, t_2, \ldots, t_S\}$ denote an observed data pattern where t_S is a subset of $\{X_1, X_2, \ldots, X_K\}$ and denotes a set of observed variables for a group of individuals. Fig. 1 describes that variables in $t_1 = \{X_1, X_2\}$ are observed for n_1 individuals in group 1, but X_3 and X_4 are missing; variables in $t_2 = \{X_2, X_3\}$ are observed for group 2, but X_1 and X_4 are missing; and variables in $t_3 = \{X_3, X_4\}$ are observed for group 3, but X_1 and X_2 are missing.

Let $Y = (Y_{\text{obs}}, Y_{\text{mis}})$ where Y_{obs} denotes the observed data for all individuals and Y_{mis} denotes the missing data for all individuals. In Fig. 2, Y_{obs} denotes all observed data and Y_{mis} denotes all missing data.

Further, let $Y_{\text{mis}} = (Y_{\text{mis}1}, Y_{\text{mis}2})$ where $Y_{\text{mis}1}$ denotes a part of missing data that will be imputed at the E step of the PIEM algorithm, and where $Y_{\text{mis}2}$ denotes the other part of the missing data that will not be imputed at the E step. In Fig. 3, a monotone data pattern is obtained by imputing missing data of X_1 for individuals in group 2 and missing data of X_1 and X_2 for group 3 (i.e., $Y_{\text{mis}1}$). If variable X_k is observed, then missing

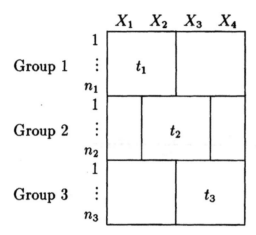

Figure 1 An observed data pattern.

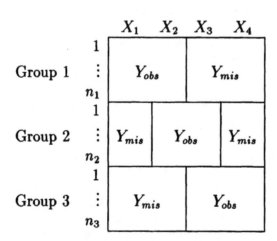

Figure 2 Incompletely observed data.

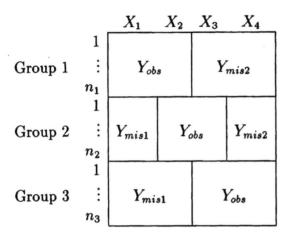

Figure 3 Monotone data pattern obtained by partial imputation.

values of variables $X_1,\ldots,X_{k-1}$ are imputed such that the observed data and imputed data construct a monotone data pattern. The likelihood function for observed data and imputed data can be factorized as

$$L\left(\phi\middle|Y_{\text{obs}}, Y_{\text{mis1}}\right) = \left[\prod_{k=1}^{3}\prod_{i\in\text{group}\,k} f(x_{1i}, x_{2i}; \phi_{12})\right]$$

$$\times \left[\prod_{k=2}^{3}\prod_{i\in\text{group}\,k} f(x_{3i}|x_{1i}, x_{2i}; \phi_{3|12})\right]$$

$$\times \left[\prod_{i\in\text{group}\,3} f(x_{4i}|x_{1i}, x_{2i}, x_{3i}; \phi_{4|123})\right],$$

where x_{ji} denotes value of X_j for individual i. If parameters ϕ_{12}, $\phi_{3|12}$, and $\phi_{4|123}$ are distinct, then each factor corresponds to a likelihood for complete data, and maximum likelihood estimates (MLEs) of these parameters can be obtained by maximizing each factor separately.

In the PIEM algorithm, a monotone data pattern may not be necessary. For example, if variables in X_1 are independent of X_3 and X_4 conditional on X_2, then we impute only missing values of X_2 for individuals in group 3 (i.e., Y'_{mis1} in Fig. 4). The likelihood function for observed data and imputed data can be factorized as

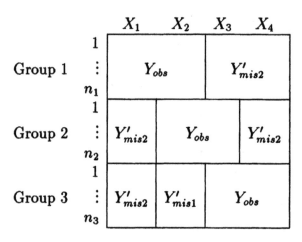

Figure 4 Nonmonotone data pattern of Y_{obs} and Y'_{mis1}.

$$L(\phi|Y_{\text{obs}}, Y'_{\text{mis1}}) = \left[\prod_{k=1}^{3} \prod_{i \in \text{group } k} f(x_{2i}; \phi_{23})\right]$$

$$\times \left[\prod_{i \in \text{group } 1} f(x_{1i}|x_{2i}; \phi_{1|2})\right]$$

$$\times \left[\prod_{k=2}^{3} \prod_{i \in \text{group } k} f(x_{3i}|x_{2i}; \phi_{3|2})\right]$$

$$\times \left[\prod_{i \in \text{group } 3} f(x_{4i}|x_{2i}, x_{3i}; \phi_{4|23})\right].$$

If parameters ϕ_2, $\phi_{1|2}$, $\phi_{3|2}$, and $\phi_{4|23}$ are distinct, then each factor corresponds to a likelihood for complete data and MLEs of these parameters can be obtained by maximizing each factor separately at the M step.

2 CONVERGENCE OF THE PIEM ALGORITHM

Let ϕ be a $1 \times d$ parameter vector. The log-likelihood for Y_{obs} is

$$l(\phi|Y_{\text{obs}}) = \log f(Y_{\text{obs}}|\phi).$$

Similar to Dempster et al. (2), we have

$$l(\phi|Y_{\text{obs}}) = Q^*\left(\phi\big|\phi^{(t)}\right) - H^*\left(\phi\big|\phi^{(t)}\right)$$

where $\phi^{(t)}$ is the current estimate of ϕ,

$$Q^*\left(\phi|\phi^{(t)}\right) = \int l^*\left(\phi|Y_{\text{obs}}, Y_{\text{mis1}}\right) f\left(Y_{\text{mis1}}\big|Y_{\text{obs}}, \phi^{(t)}\right) dY_{\text{mis1}}$$

and

$$H^*\left(\phi|\phi^{(t)}\right) = \int \left[\log f(Y_{\text{mis1}}|Y_{\text{obs}}, \phi)\right] f\left(Y_{\text{mis1}}\big|Y_{\text{obs}}, \phi^{(t)}\right) dY_{\text{mis1}},$$

where $l^*(\phi|Y_{\text{obs}}, Y_{\text{mis1}})$ is the log-likelihood for the observed data and the imputed data ($Y_{\text{obs}}, Y_{\text{mis1}}$).

For exponential families, at the E step of the PIEM algorithm, Y_{mis1} need not be imputed and only the sufficient statistic $s(Y_{\text{obs}}, Y_{\text{mis1}})$ is estimated by

$$s^{(t+1)} = E\left(s(Y_{\text{obs}}, Y_{\text{mis1}})\big|Y_{\text{obs}}, \phi^{(t)}\right).$$

At the M step of the PIEM algorithm, the likelihood is factored based on the pattern of imputed data,

$$l^*(\phi|Y_{\text{obs}}, Y_{\text{mis1}}) = \sum_i l_i^*(\phi_i|Y_{\text{obs}}, Y_{\text{mis1}}),$$

so that parameters $\phi_1, \ldots, \phi_I$ are distinct and each factor $l^*_i(\phi_i|Y_{\text{obs}}, Y_{\text{mis1}})$ corresponds to a log-likelihood for a complete data problem. Thus

$$Q^*\left(\phi|\phi^{(t)}\right) = \int l^*\left(\phi|Y_{\text{obs}}, Y_{\text{mis1}}\right) f\left(Y_{\text{mis1}}\big|Y_{\text{obs}}, \phi^{(t)}\right) dY_{\text{mis1}}$$

$$= \sum_i \int l_i^*\left(\phi_i|Y_{\text{obs}}, Y_{\text{mis1}}\right) f\left(Y_{\text{mis1}}\big|Y_{\text{obs}}, \phi^{(t)}\right) dY_{\text{mis1}}$$

$$= \sum_i Q_i^*\left(\phi_i|\phi^{(t)}\right),$$

where $Q^*_i(\phi_i|\phi^{(t)}) = \int l^*_i(\phi_i|Y_{\text{obs}}, Y_{\text{mis1}}) f(Y_{\text{mis1}}|Y_{\text{obs}}, \phi^{(t)}) dY_{\text{mis1}}$. Thus $Q^*(\phi|\phi^{(t)})$ can be maximized by maximizing each component $Q^*_i(\phi_i|\phi^{(t)})$ separately.

Theorem 1. *At each iteration, the PIEM algorithm monotonically increases the log-likelihood function* $l(\phi \mid Y_{obs})$, *that is,*

$$l\left(\phi^{(t+1)} \middle| Y_{obs}\right) \geq l\left(\phi^{(t)} \middle| Y_{obs}\right).$$

Proof. By Jensen's inequality, we have

$$H^*\left(\phi \middle| \phi^{(t)}\right) \leq H^*\left(\phi^{(t)} \middle| \phi^{(t)}\right).$$

Thus

$$l\left(\phi^{(t+1)} \middle| Y_{obs}\right) - l\left(\phi^{(t)} \middle| Y_{obs}\right) = \left[Q^*\left(\phi^{(t+1)} \middle| \phi^{(t)}\right) - Q^*\left(\phi^{(t)} \middle| \phi^{(t)}\right)\right]$$

$$- \left[H^*\left(\phi^{(t+1)} \middle| \phi^{(t)}\right) - H^*\left(\phi^{(t)} \middle| \phi^{(t)}\right)\right] \geq 0.$$

Theorem 1 means that $l(\phi \mid Y_{obs})$ is nondecreasing in each iteration of the PIEM algorithm, and that it is strictly increasing when $Q^*(\phi^{(t+1)} \mid \phi^{(t)}) > Q^*(\phi^{(t)} \mid \phi^{(t)})$. The PIEM algorithm has the same general conditions for convergence as those for the EM algorithm, that is, $l(\phi \mid Y_{obs})$ is bounded above (2,5). Now we compare the convergence rate of the PIEM algorithm with that of the ordinary EM algorithm and show that the PIEM algorithm has a faster rate of convergence than the EM algorithm. For the PIEM algorithm, let $M^*(\phi)$ be a mapping from the parameter space to itself so that each step of the PIEM algorithm $\phi^{(t)} \to \phi^{(t+1)}$ is defined by $\phi^{(t+1)} = M^*(\phi^{(t)})$, for $t = 0, 1, \ldots$. Assume that $M^*(\phi)$ is differentiable at $\hat{\phi}$ and that $\phi^{(t)}$ converges to the MLE $\hat{\phi}$, that is, $M^*(\hat{\phi}) = \hat{\phi}$. Denote

$$\mathrm{DM}^*(\phi) = \left(\frac{\partial M_j^*(\phi)}{\partial \phi_i}\right),$$

where $M^*(\phi) = (M_1^*(\phi), \ldots, M_d^*(\phi))$. Applying Taylor expansion to $M^*(\phi^{(t)})$ at $\hat{\phi}$, we get

$$\phi^{(t+1)} - \hat{\phi} = \left(\phi^{(t)} - \hat{\phi}\right)\mathrm{DM}^*\left(\hat{\phi}\right) + O\left(\left\| \phi^{(t)} - \hat{\phi} \right\|^2\right).$$

The rate of convergence is defined as

$$R^* = \lim_{t \to \infty} \frac{\left\| M^*\left(\phi^{(t)}\right) - \hat{\phi} \right\|}{\left\| \phi^{(t)} - \hat{\phi} \right\|},$$

where $\|\cdot\|$ is the Euclidean norm. Thus the rate of convergence of the PIEM algorithm is the largest eigenvalue of $DM^*(\hat{\phi})$, as discussed in Refs. 2 and 4.

For the EM algorithm, Dempster et al. (2) showed that

$$l(\phi|Y_{obs}) = Q\left(\phi|\phi^{(t)}\right) - H\left(\phi|\phi^{(t)}\right)$$

where

$$Q\left(\phi|\phi^{(t)}\right) = \int l\left(\phi|Y_{obs}, Y_{mis}\right) f\left(Y_{mis}|Y_{obs}, \phi^{(t)}\right) dY_{mis}$$

and

$$H\left(\phi|\phi^{(t)}\right) = \int \left[\log f\left(Y_{mis}|Y_{obs}, \phi\right)\right] f\left(Y_{mis}|Y_{obs}, \phi^{(t)}\right) dY_{mis}.$$

Let $M(\phi)$ be the mapping $\phi^{(t)} \to \phi^{(t+1)}$ for the EM algorithm and R as its rate of convergence:

$$R = \lim_{t \to \infty} \frac{\left\| M\left(\phi^{(t)}\right) - \hat{\phi} \right\|}{\left\| \phi^{(t)} - \hat{\phi} \right\|}.$$

Let D^{20} denote the second derivative with respect to the first argument, so that

$$D^{20}Q^*(\phi''|\phi') = \frac{\partial^2}{\partial\phi \cdot \partial\phi} Q^*(\phi|\phi')|_{\phi=\phi''},$$

and define

$$I_c^*(\phi|Y_{obs}) = -D^{20}Q^*(\phi|\phi), \quad I_c(\phi|Y_{obs}) = -D^{20}Q(\phi|\phi),$$
$$I_m^*(\phi|Y_{obs}) = -D^{20}H^*(\phi|\phi), \quad I_m(\phi|Y_{obs}) = -D^{20}H(\phi|\phi),$$

and

$$I_o^*(\phi|Y_{obs}) = -D^{20}l(\phi|Y_{obs}).$$

Then $I_o^* = I_c^* - I_m^*$. We can show that I_c^*, I_c, I_m^*, I_m, I_o^*, and $I_c - I_c^*$ are symmetric and nonnegative definite.

To prove that the convergence of the PIEM algorithm is faster than the EM algorithm, we first show a lemma about eigenvalues of matrices.

Let $\lambda_1[A]$ be the largest eigenvalue of a matrix A, and let A^T denote the transpose of A.

Lemma 1. *If B is a symmetric and positive definite matrix and if A, C, and $B-A$ are symmetric and nonnegative definite matrices, then $\lambda_1[(A+C)(B+C)^{-1}]\geq\lambda_1[AB^{-1}]$ with strict inequality if C and $B-A$ are positive definite.*

Proof. Since B is symmetric and positive definite, we can write $B^{-1}=B^{-1/2}B^{-1/2}$. From Proposition 1.39 of Eaton (3, p. 48), we get

$$\lambda_1\left(AB^{-1}\right) = \lambda_1\left(AB^{-\frac{1}{2}}B^{-\frac{1}{2}}\right) = \lambda_1\left(B^{-\frac{1}{2}}AB^{-\frac{1}{2}}\right).$$

Since $B^{-1/2}AB^{-1/2}$ is a Grammian matrix (that is, a symmetric and non-negative definite matrix), we have from Theorem 5.24 of Basilevsky (1983)

$$\lambda_1\left(AB^{-1}\right) = \lambda_1\left(B^{-\frac{1}{2}}AB^{-\frac{1}{2}}\right) = \max_{x\neq0}\left(\frac{x^T B^{-\frac{1}{2}}AB^{-\frac{1}{2}}x}{x^T x}\right).$$

Let $y=B^{-1/2}x$. Then

$$\lambda_1\left(AB^{-1}\right) = \max_{y\neq0}\left(\frac{y^T Ay}{y^T By}\right).$$

Substituting $A+C$ for A and $B+C$ for B, we also get

$$\lambda_1\left[(A+C)(B+C)^{-1}\right] = \max_{z\neq0}\frac{z^T(A+C)z}{z^T(B+C)z}.$$

It is clear that for any nonnegative constants a, b, and c where $a\leq b$ and $b>0$, we have $(a+c)/(b+c)\geq a/b$. Thus we obtain, for any vector $y\neq0$,

$$\frac{y^T(A+C)y}{y^T(B+C)y} = \frac{y^T Ay + y^T Cy}{y^T By + y^T Cy} \geq \frac{y^T Ay}{y^T By}.$$

Therefore $\lambda_1[(A+C)(B+C)^{-1}]\geq\lambda_1[AB^{-1}]$ with strict inequality if C and $B-A$ are positive definite.$\square$

Theorem 2. *Suppose that $I_c^*(\hat{\phi}|Y_{obs})$ is positive definite. Then the PIEM algorithm converges faster than the EM algorithm, that is, $R^*\leq R$ with strict inequality if $I_o^*(\hat{\phi}|Y_{obs})$ and $I_c(\hat{\phi}|Y_{obs})-I_c^*(\hat{\phi}|Y_{obs})$ are positive definite.*

Proof. We can show easily that

$$Q(\phi|\phi') - Q^*(\phi|\phi') = H(\phi|\phi') - H^*(\phi|\phi') = \Delta(\phi|\phi'),$$

where

$$\Delta(\phi|\phi') = \int [\log f(Y_{\text{mis2}}|Y_{\text{obs}}, Y_{\text{mis1}}, \phi)] f(Y_{\text{mis}}|Y_{\text{obs}}, \phi') dY_{\text{mis}}.$$

From the definitions of I_{m}^*, I_{m}, I_{c}^*, and I_{c}, we can obtain

$$I_{\text{m}}^*(\hat{\phi}|Y_{\text{obs}}) = I_{\text{m}}(\hat{\phi}|Y_{\text{obs}}) + D^{20}\Delta\left(\hat{\phi}|\hat{\phi}\right)$$

and

$$I_{\text{c}}^*(\hat{\phi}|Y_{\text{obs}}) = I_{\text{c}}(\hat{\phi}|Y_{\text{obs}}) + D^{20}\Delta\left(\hat{\phi}|\hat{\phi}\right).$$

$D^{20}\Delta(\hat{\phi}|\hat{\phi})$ is nonpositive definite since

$$D^{20}\Delta\left(\hat{\phi}|\hat{\phi}\right) = \frac{\partial^2}{\partial\phi^2} E\left[\log f(Y_{\text{mis2}}|Y_{\text{obs}}, Y_{\text{mis1}}, \phi)\bigg| Y_{\text{obs}}, \hat{\phi}\right]_{\phi=\hat{\phi}}$$

$$= E\left[\frac{\partial^2}{\partial\phi^2} \log f(Y_{\text{mis2}}|Y_{\text{obs}}, Y_{\text{mis1}}, \phi)\bigg|_{\phi=\hat{\phi}}\bigg| Y_{\text{obs}}, \hat{\phi}\right]$$

$$= -E\left[\left\{\frac{\partial}{\partial\phi} \log f(Y_{\text{mis2}}|Y_{\text{obs}}, Y_{\text{mis1}}, \phi)\bigg|_{\phi=\hat{\phi}}\right\}\right.$$

$$\left. \times \left\{\frac{\partial}{\partial\phi} \log f(Y_{\text{mis2}}|Y_{\text{obs}}, Y_{\text{mis1}}, \phi)\bigg|_{\phi=\hat{\phi}}\right\}^{\text{T}}\bigg| Y_{\text{obs}}, \hat{\phi}\right],$$

where A^{T} denotes the transpose of matrix A. From Ref. 2, we get

$$\text{DM}\left(\hat{\phi}\right) = I_{\text{m}}\left(\hat{\phi}|Y_{\text{obs}}\right) I_{\text{c}}\left(\hat{\phi}|Y_{\text{obs}}\right)^{-1}.$$

Similarly, we have

$$\text{DM}^*\left(\hat{\phi}\right) = I_{\text{m}}^*\left(\hat{\phi}|Y_{\text{obs}}\right) I_{\text{c}}^*\left(\hat{\phi}|Y_{\text{obs}}\right)^{-1}.$$

Thus

$$\text{DM}\left(\hat{\phi}\right) = \left[I_{\text{m}}^*\left(\hat{\phi}|Y_{\text{obs}}\right) - D^{20}\Delta\left(\hat{\phi}|\hat{\phi}\right)\right]$$

$$\times \left[I_{\text{c}}^*\left(\hat{\phi}|Y_{\text{obs}}\right) - D^{20}\Delta\left(\hat{\phi}|\hat{\phi}\right)\right]^{-1}.$$

Since $-D^{20}\Delta(\hat{\phi}|\hat{\phi})$ and $I_{\text{o}}^* = I_{\text{c}}^* - I_{\text{m}}^*$ are symmetric and nonnegative definite, we have from Lemma 1 that $\lambda_1[\text{DM}^*(\hat{\phi})] \leq \lambda_1[\text{DM}(\hat{\phi})]$, that is, $R^* \leq R$. If $-D^{20}\Delta(\hat{\phi}|\hat{\phi})$ and I_{o}^* are positive definite, then $R^* < R$.

Let $(Y_{\text{obs}}, Y_{\text{mis1}})$ and $(Y_{\text{obs}}, Y'_{\text{mis1}})$ be two different patterns of imputed data so that Y'_{mis1} is contained in Y_{mis1}, that is, $(Y_{\text{obs}}, Y'_{\text{mis1}})$ has fewer imputed data than (Y_{obs}, Y_{mis1}). For example, the imputed missing data Y'_{mis1} in Fig. 4 are less than those Y_{mis1} in Fig. 3. In a similar way, we can show that the PIEM algorithm imputing Y'_{mis1} converges faster than that imputing Y_{mis1}. This means that imputing fewer variables is preferable. The less the imputed data are, the faster the convergence is.

3 EXAMPLES

In this section, we give several numerical examples to illustrate the PIEM algorithm and its convergence.

Example 1. Let X_1 and X_2 be binary variables. Suppose that X_1 and X_2 follow multinomial distribution with parameters $p(ij) = \Pr(X_1 = i, X_2 = j)$ and the sample size n. Suppose that the observed data pattern is $T = \{\{X_1, X_2\}, \{X_1\}, \{X_2\}\}$. The observed data are shown in Table 1 and sample size $n = 520$. $n_{12}(ij)$'s, $n_1(i)$'s, and $n_2(j)$'s are observed frequencies where the subscript A of n_A denotes an index set of observed variables. $n_{12}(ij)$'s denote completely observed data, and $n_1(i)$'s denote incompletely observed data for which X_1 is observed but X_2 is missing. From Table 1, we can see that X_2 is missing much more often than X_1. The MLEs $\hat{p}(ij)$'s of cell probabilities are shown in Table 2, and the numbers of iterations for the EM and PIEM algorithms are shown in Table 3.

As shown in Table 3, there are two different methods for imputing incomplete data. The first method, (a) "$n_2(j) \rightarrow n_2(ij)$," is to impute $n_2(j)$ to $n_2(ij)$ as follows:

- the E step:

$$\hat{n}_2(ij) = \frac{\hat{p}^{(t)}(ij)}{\hat{p}^{(t)}(+j)} n_2(j);$$

Table 1 Observed Data

$n_{12}(ij)$	$j = 1$	$j = 2$	$n_1(i)$
$i = 1$	5	4	300
$i = 2$	2	1	200
$n_2(j)$	5	3	

Table 2 MLEs of parameters p_{ij}

$\hat{p}(ij)$	$j=1$	$j=2$
$i=1$	0.3402	0.2633
$i=2$	0.2700	0.1265

- the M step:

$$\hat{p}^{(t+1)}(ij) = \frac{n_{12}(ij) + \hat{n}_2(ij)}{n_{12}(i+) + \hat{n}_2(i+)} \cdot \frac{n_{12}(i+) + n_1(i) + \hat{n}_2(i+)}{n_{12}(++) + n_1(+) + n_2(+)}.$$

The second method, (b) "$n_1(i) \rightarrow n_1(ij)$," is to impute $n_1(i)$ to $n_1(ij)$. We can see that both of the PIEM algorithms converge faster than the EM algorithm. The first method converges faster than the second one. A main reason is that the former imputes less missing data than the latter, but it is not always true since the imputed data of the former are not a subset of the latter's imputed data.

Example 2. Let $(X_1, X_2)^T$ be a bivariate random vector following a normal distribution $N(\mu, \Sigma)$. Suppose that the observed data pattern is $T = \{\{X_1, X_2\}, \{X_1\}, \{X_2\}\}$. The observed data are shown in Table 4.

The EM algorithm imputes all missing data of incompletely observed data. From the properties of the normal distribution, the conditional distribution of X_i given $X_j = x_j$ is also a normal distribution, and its mean and variance are, respectively,

$$\mu_i + \frac{\sigma_{ij}}{\sigma_{jj}} (x_j - \mu_j)$$

and

$$\sigma_{ii.j} = \sigma_{ii}(1 - \rho^2),$$

Table 3 Numbers of Iteration ($|\hat{p}^{(t+1)}(ij) - \hat{p}^{(t)}(ij)| \leq 10^{-5}$)

Algorithm	Iteration number
EM	253
PIEM $n_2(j) \rightarrow n_2(ij)$	11
PIEM $n_1(i) \rightarrow n_1(ij)$	251

Table 4 Observed Data

t_s	Individual	X_1	X_2
$\{X_1,X_2\}$	1	1.2	2.3
	2	1.7	0.1
	3	1.6	−0.7
$\{X_1\}$	4	0.2	?
	5	1.5	?
$\{X_2\}$	6	?	−0.2
	7	?	1.6

where μ_i is the mean of X_i, σ_{ij} is the covariance of X_i and X_j, and ρ is the correlation of X_1 and X_2. Thus the missing value of X_i of the individual with an observed value $X_j = x_j$ is imputed by

$$\hat{x}_i = \mu_i^{(k)} + \frac{\sigma_{ij}^{(k)}}{\sigma_{jj}^{(k)}}\left(x_j - \mu_j^{(k)}\right),$$

where $\sigma_{ij}^{(k)}$ and $\mu_i^{(k)}$ denote the estimates obtained at the kth iteration of the EM algorithm. The PIEM algorithm imputes only a part of missing data such that the observed data and imputed data construct a monotone pattern. There are two ways for imputation: one is imputing the missing data of the incomplete data with the observed variate set $\{X_1\}$, and the other is imputing those with the observed variable set $\{X_2\}$. The same MLEs of μ and Σ are obtained by using the EM algorithm and the PIEM algorithm as follows:

$$\hat{\mu} = \begin{pmatrix} 1.3005 \\ 1.4163 \end{pmatrix}, \ \hat{\Sigma} = \begin{pmatrix} 0.2371 & -1.0478 \\ -1.0478 & 4.9603 \end{pmatrix}.$$

Table 5 Convergence of Algorithms

Algorithm	Iterations	Rate of convergence
EM	282	0.9526
PIEM (Impute $\{X_1\}$ to complete)	53	0.7075
PIEM (Impute $\{X_2\}$ to complete)	250	0.9474

The convergence of the EM algorithm and the two PIEM algorithms are shown in Table 5. Both of the PIEM algorithms converge faster than the EM algorithm. The first PIEM algorithm treats values of X_2 for individuals 4 and 5 as missing data, but the second PIEM algorithm treats values of X_1 for individuals 6 and 7 as missing data. From the MLE of covariance matrix, it can be seen that the variance of X_1 is much less than that of X_2. Thus the first PIEM algorithm has less missing information than the second algorithm, and then the first one converges faster than the second one.

REFERENCES

1. Basilevsky, A. (1983). *Applied Matrix Algebra in the Statistical Sciences*. New York: North-Holland.
2. Dempster, A. P., Laird, N. M., Rubin, D. B. (1977). Maximum likelihood estimation from incomplete data via the EM algorithm (with discussion). *J. R. Stat. Soc. Ser. B* 39:1–38.
3. Eaton, M. L. (1983). *Multivariate Statistics: A Vector Space Approach*. New York: Wiley.
4. Meng, X. L. (1994). On the rate of convergence of the ECM algorithm. *Ann. Stat.* 22:326–339.
5. Wu, C. F. J. (1983). On the convergence properties of the EM algorithm. *Ann. Stat.* 11:95–103.

7

Convergence Speed and Acceleration of the EM Algorithm

Mihoko Minami
The Institute of Statistical Mathematics, Tokyo, Japan

1 INTRODUCTION

An attractive feature of the EM algorithm is its simplicity. It is often used as an alternative to the Newton–Raphson method and other optimization methods when the latter are too complicated to implement. However, it is often criticized that the convergence of the EM algorithm is too slow. Whether its slow convergence is a real problem in practice does depend on models, data sizes, and situation programs used. Many acceleration methods have been proposed to speed up the convergence of the EM algorithm. Also, hybrid methods that would switch from the EM algorithm to other optimization algorithm with faster convergence rate have been proposed.

This chapter discusses the convergence speed of the EM algorithm and other iterative optimization methods and reviews the acceleration methods of the EM algorithm for finding the maximum likelihood estimate.

2 CONVERGENCE SPEED

We start from the definition of convergence speed and convergence rate of iterative numerical methods. We denote the vector of parameter values after the kth iteration by $\theta^{(k)}$ and its converged point by θ^*. An iterative

numerical method is said to converge linearly if it holds that with some
constant c $(0 < c < 1)$ and positive integer k_0,

$$\|\theta^{(k+1)} - \theta^*\| \leq c\|\theta^{(k)} - \theta^*\| \quad \text{for any } k \geq k_0. \tag{1}$$

The constant c is called the convergence rate. If it holds that with some
sequence $\{c_k\}$ converging to 0 and positive integer k_0,

$$\|\theta^{(k+1)} - \theta^*\| \leq c_k\|\theta^{(k)} - \theta^*\| \quad \text{for any } k \geq k_0, \tag{2}$$

then the method is said to converge superlinearly. If it holds that with some
constant c $(0 < c < 1)$ and positive integer k_0,

$$\|\theta^{(k+1)} - \theta^*\| \leq c\|\theta^{(k)} - \theta^*\|^2 \quad \text{for any } k \geq k_0, \tag{3}$$

then the method is said to converge quadratically. A numerical method
with the superlinear or quadratic convergence property converges rapidly
after the parameter value comes close to θ^*, while a method with the linear
convergence property might take fairly large number of iterations even
after the parameter value comes close to θ^*.

Hereafter, we denote the incomplete data by y, the incomplete-data
log-likelihood by $l(\theta; y)$, the observed incomplete-data information matrix
by $I(\theta; y)$, and the expected complete-data information matrix by $\mathcal{I}_c(\theta; y)$.

The EM update can be approximated by (cf. Ref. 1):

$$\text{EM update}: \quad \theta^{(k+1)} \approx \theta^{(k)} + \mathcal{I}_c^{-1}\left(\theta^{(k)}; y\right) \frac{\partial l}{\partial \theta}\left(\theta^{(k)}; y\right). \tag{4}$$

When the complete data belong to the regular exponential family and the
mean parameterization is employed, the above approximation becomes
exact. It is further approximated by

$$\theta^{(k+1)} \approx \theta^{(k)} - \mathcal{I}_c^{-1}\left(\theta^{(k)}; y\right) I(\theta^*; y)\left(\theta^{(k)} - \theta^*\right)$$

$$\theta^{(k+1)} - \theta^* \approx \left(E - \mathcal{I}_c^{-1}\left(\theta^{(k)}; y\right) I(\theta^*; y)\right)\left(\theta^{(k)} - \theta^*\right)$$

where E denotes the identity matrix. Thus the EM algorithm converges
linearly and its convergence rate is the largest eigenvalue of $E - \mathcal{I}_c^{-1}(\theta^*; y)$
$I(\theta^*; y)$, that is, $1 - \gamma$ where γ is the smallest eigenvalue of $\mathcal{I}_c^{-1}(\theta^*; y) I(\theta^*; y)$.
Meng and van Dyk (2) consider speeding up the convergence of the EM
algorithm by introducing a working parameter in their specification of the
complete data. Their idea is to find the optimal data augmentation in a
sense that γ is maximized within the specified type of data augmentation.
Their method speeds up the convergence within the framework of the basic

EM algorithm without affecting its stability and simplicity. However, the resulting EM algorithm still converges linearly. Moreover, such complete-data specification might not exist in general.

The Newton–Raphson method approximates the objective function (the incomplete-data log-likelihood function) by a quadratic function and takes its maximizer as the next parameter value. Its update formula is:

Newton – Raphson update :

$$\theta^{(k+1)} = \theta^{(k)} + I^{-1}\left(\theta^{(k)};y\right)\frac{\partial l}{\partial \theta}\left(\theta^{(k)};y\right). \tag{5}$$

The Newton–Raphson method converges quadratically. The observed information matrix $I(\theta^{(k)};y)$ might not be positive definite, especially at the beginning of iteration. In order to obtain a broader range of convergence, one ought to modify it to be positive definite if it is not. Moreover, a line search as in the case of quasi-Newton methods described below would be better to be employed (3).

Quasi-Newton methods do not directly compute the observed information matrix $I(\theta;y)$ but approximate it in the direction of convergence. Its update formula is

$$\text{Quasi – Newton update :} \quad \theta^{(k+1)} = \theta^{(k)} + \alpha_k B_k^{-1}\frac{\partial l}{\partial \theta}\left(\theta^{(k)};y\right) \tag{6}$$

where matrix B_k is updated using only the change in gradient $q_k = \partial l/\partial\theta(\theta^{(k)};y) - \partial l/\partial\theta(\theta^{(k-1)};y)$ and the change in parameter value $s_k = \theta^{(k)} - \theta^{(k-1)}$. The most popular quasi-Newton update is the Broyden–Fletcher–Goldfarb–Shanno (BFGS) update:

$$B_{k+1} = B_k - \frac{B_k s_k s_k^T B_k}{s_k^T B_k s_k} - \frac{q_k q_k^T}{q_k^T s_k}. \tag{7}$$

Step length α_k is determined by a line search iteration so that it satisfies the conditions that ensure the convergence of iterates, such as the Wolfe conditions:

$$l\left(\theta^{(k)} + \alpha_k d_k\right) \geq l\left(\theta^{(k)}\right) + \mu\alpha_k\frac{\partial l}{\partial\theta}\left(\theta^{(k)};y\right)^T d_k$$

$$\frac{\partial l}{\partial\theta}\left(\theta^{(k)} + \alpha_k d_k;y\right)^T d_k \leq \eta\frac{\partial l}{\partial\theta}\left(\theta^{(k)};y\right)^T d_k$$

where $d_k = B_k^{-1}\partial l/\partial\theta(\theta^{(k)};y)$, $0 < \mu \leq 0.5$, and $\mu \leq \eta < 1$. Typical values in practice are $\mu = 10^{-4}$ and $\eta = 0.5$. As algorithms converge, $\alpha = 1$ is getting to satisfy the conditions. The convergence speed of quasi-Newton algorithms is superlinear. The conjugate gradient method is another well-known iterative method with superlinear convergence. For more theoretical details for the above numerical methods, see Refs. 4–6.

If we compare simply the numbers of iterations until algorithms converge, the Newton–Raphson method would take the fewest iterations. However, the Newton–Raphson method requires the observed information matrix, i.e., the Hessian of the objective function takes more computational time than the EM update or the gradient of the objective function. Lindstrom and Bates (3) compared the number of iterations and the computational time of the EM algorithm and the Newton–Raphson method for mixed linear model. They reported that the Newton–Raphson method is as stable as the EM algorithm if the observed information matrix is modified to be positive definite and a line search is employed. Computational time per iteration for the Newton–Raphson method takes as three or four times as the EM algorithm, but the number of iterations is fewer than 1/10 as that for the EM algorithm. For some dataset, the Newton–Raphson converged after six iterations, while the EM algorithm did not converge even after 200 iterations.

An attractive feature of the quasi-Newton methods is that we need not compute the observed information matrix. Update of B_k is typically of low rank (1 or 2) and requires quite small computational time. Since the computational complexity of the gradient of the incomplete-data log-likelihood function is not much different from that for the EM update in many models, the computational time per iteration for quasi-Newton methods is not much more than that for the EM algorithm. It is often criticized that the quasi-Newton methods perform poorly at the beginning of iterations. This can be avoided if one uses the expected complete-data information matrix $\mathcal{I}_c(\theta)$ as the initial value B_0 (7). Then, as Eq. (4) suggests, the first iteration of the quasi-Newton method is approximately (exactly with mean parameterization) equal to the EM update. In addition, one can use the EM algorithm for the first several iterations to take advantage of its good global convergence properties, and then switch to quasi-Newton methods.

3 COMPARISON OF COMPUTING TIME

In this section, we compare the iteration numbers and computing time by the EM algorithm and the quasi-Newton method with the BFGS update

for a mixed linear model and a Gaussian mixture model. For the quasi-Newton method, we used the EM algorithm for the first three iterations and switched to the quasi-Newton method with $\mathcal{I}_c\left(\theta^{(3)}\right)$ as the initial value B_0.

3.1 A Mixed Linear Model

We consider the following mixed linear model:

$$y = X\beta + Za + e \tag{8}$$

where y is the $n \times 1$ vector of observations, X and Z are known design matrices, β is the $p \times 1$ vector of fixed effects, a is the $q \times 1$ vector of random effects, and e is the $n \times 1$ vector of errors. Random effects a and errors e are assumed to follow normal distributions with mean zero. In the field of animal breeding, mixed linear models are often used to accommodate genetical relationship among individual animals. In the individual animal model (IAM) by Saito and Iwasaki (8), the covariance matrix of random effects a is assumed to be expressed as $\sigma_a^2 A$ where matrix A is obtained from the mating structure of animals, and the covariance matrix of e as $\sigma_e^2 I$, that is, $a \sim N(0,\sigma_a^2 A)$ and $e \sim N(0,\sigma_e^2 I)$. Observations y are, for example, amounts of milk that individual cows produced in a day, β are effects of years and cowsheds, and a are effects of individual cows. For the estimation of variance components, the restricted maximum likelihood (REML) estimation is often used. The REML estimation finds the estimate by maximizing error contrast $M^T y$ where M is a matrix of size $n \times (n-p)$ with full column rank and is orthogonal to X. The REML estimation takes into account the loss in degrees of freedom by estimating fixed effects. See Ref. 9 for the REML estimation by the EM algorithm for mixed linear models.

Table 1 shows computational times for the REML estimation of variance components σ_a^2 and σ_e^2 by the EM algorithm and the quasi-Newton method with BFGS update for the data given in Ref. 8. For this data, $n = q = 14$ and $p = 2$. Computing times are the averages of 1000 repetitions. Algorithms were terminated when the relative change in the estimates or the incomplete-data log-likelihood became less than eps $= 10^{-6}$, 10^{-7}, or 10^{-8}.

The quasi-Newton method converged much faster than the EM algorithm with much fewer iterations. The quasi-Newton method took less than 1/10 in computational time and iteration numbers of the EM

Table 1 Computing Time of the EM Algorithm and a Quasi-Newton Method for a Mixed Linear Model

Method	eps	Iter.	Time	T/I	$\hat{\sigma}_a^2$	$\hat{\sigma}_e^2$
EM	10^{-6}	220	1.0191	0.00463	0.64172922	0.73217675
	10^{-7}	280	1.2941	0.00462	0.64171389	0.73218902
	10^{-8}	340	1.5737	0.00463	0.64171236	0.73219024
QN	10^{-6}	19	0.0916	0.00482	0.64171219	0.73219038
	$10^{-7}, 10^{-8}$	21	0.0998	0.00475	0.64171219	0.73219038

From the left: eps—stopping criteria, iter.—number of iterations, time—total computing time, T/I—computing time per iteration, and $\hat{\sigma}^2_a$ and $\hat{\sigma}^2_e$—estimates of the variances.

algorithm. Computational time per iteration is almost the same for both methods. This is because the computational complexity of the derivatives of the log-likelihood function is almost the same as that of the EM update. The derivatives of the log-likelihood function are expressed with the EM map for this model:

$$\frac{\partial L(\theta)}{\partial \sigma_a^2} = \frac{q}{2\sigma_a^4} \left(M_a(\theta) - \sigma_a^2 \right)$$

$$\frac{\partial L(\theta)}{\partial \sigma_e^2} = \frac{n}{2\sigma_e^4} \left(M_e(\theta) - \sigma_e^2 \right)$$

where $M_a(\theta)$ and $M_e(\theta)$ denote the EM map for σ_a^2 and σ_e^2, respectively. The convergence rate of the EM algorithm for these data is 0.96. It should be mentioned that for iterative algorithms with linear convergence, the small relative change in parameter or the log-likelihood does not imply the convergence of algorithms. The estimate σ_a^2 by the EM algorithm with eps $= 10^{-6}$ matches only four digits to the one with eps $= 10^{-8}$.

3.2 Gaussian Mixture Model

The second example is a Gaussian mixture model which was also taken as an example in Dempster et al. (10). We consider a mixture model with two normal distributions with unknown mean and variance. The mixing proportion is also assumed to be unknown. Table 2 shows the computational results for the ash content data given in Ref. 11 and randomly generated 10 datasets. Among randomly generated datasets, 50 random values were generated for each distribution for the first 5 datasets, and 200

Table 2 Computing Time by the EM Algorithm and the Quasi-Newton Method for Gaussian Mixture Models

No.	Data	EM algorithm			Quasi-Newton method		
		Iter.	Time	T/I	Iter.	Time	T/I
1	ash content($n=430$)	42	0.1369	0.00326	11	0.0386	0.00351
2	50 $N(0,1)+50\ N(2,4)$	89	0.0712	0.00080	22	0.0386	0.00103
3	50 $N(0,1)+50\ N(2,9)$	137	0.1082	0.00079	19	0.0202	0.00106
4	50 $N(0,1)+50\ N(2,16)$	48	0.0385	0.00080	13	0.0132	0.00102
5	50 $N(0,1)+50\ N(1,4)$	1000 <	–	–	49	0.0476	0.00097
6	50 $N(0,5)+50\ N(2,5)$	1000 <	–	–	54	0.0548	0.00101
7	200 $N(0,1)+200\ N(2,4)$	89	0.2740	0.00308	16	0.0531	0.00332
8	200 $N(0,1)+200\ N(2,9)$	58	0.1785	0.00308	17	0.0574	0.00338
9	200 $N(0,1)+200\ N(2,16)$	61	0.1875	0.00307	17	0.0556	0.00327
10	200 $N(0,1)+200\ N(1,4)$	151	0.4610	0.00305	19	0.0632	0.00333
11	200 $N(0,5)+200\ N(2,5)$	1000 <	–	–	27	0.0885	0.00328

Iter.—number of iterations, time—total computing time, and T/I—computing time per iteration.

random values each for the rest. Programs were terminated when the relative change in the estimates or the incomplete-data log-likelihood became less than 10^{-9}. The program used double precision for real values.

In spite of the rather strict stopping criteria, the EM algorithm converges well except for datasets 5, 6, and 11. For finite mixture models, it is known that the EM algorithm performs well when the distributions are well separated (12). For datasets 5, 6, and 11, the EM algorithm did not satisfy the stopping criteria even after 1000 iterations, while the quasi-Newton method satisfied it after 49, 54, and 27 iterations, respectively. When eps = 10^{-6}, 10^{-7}, and 10^{-8}, the EM algorithm was terminated after 29, 56, and 101 iterations for dataset 5. The log-likelihood values were -102.40520, -102.39627, and -102.39481, respectively, whereas the log-likelihood value after 1000 iterations was -102.38939. The EM algorithm increases the log-likelihood value, but quite slowly for this dataset.

4 ACCELERATION METHODS

Various authors have proposed methods to speed up the EM algorithm. Jamshidian and Jennrich (13) classify them into three groups: pure, hybrid, and EM-type accelerators.

Pure accelerators are those that require only the EM algorithm for their implementation. They include the Aitkin's acceleration method (1,10,14) and methods with step lengthening (14). The Aitkin accelerator uses a generalized secant approximation to the derivative of the EM map. The Aitkin's method and its modifications (15) seem to have a problem in stability and fail to converge fairly often (3,15). Methods of step lengthening do not change the direction by the EM algorithm, but lengthen the step size when the convergence is slow. They seem to give only small gains over the EM algorithm compared with the methods that also change the direction. Jamshidian and Jennrich (13) proposed a pure accelerator QN1 that uses quasi-Newton methods to find a zero of the EM step. QN1 is simple to implement and accelerates the EM algorithm well in their numerical examples, although it is not globally convergent. In short, pure acceleration methods are relatively simple and easy to implement, but they might have a problem in stability and they are not globally convergent.

Hybrid accelerators require the EM algorithm but also use other problem-specific quantities such as the log-likelihood and its gradient. The conjugate gradient acceleration for the EM algorithm (16) and Jamshidian and Jennrich's quasi-Newton acceleration method QN2 (13) are of this type. Both algorithms require the EM step and the gradient of the incomplete-data log-likelihood, but do not compute or directly approximate the Hessian matrix. They use a line search and it makes them globally convergent like the EM algorithm. Jamshidian and Jennrich (13) showed that both methods accelerate the EM algorithm considerably well. The conjugate gradient acceleration for the EM algorithm is especially attractive when the number of parameters is very large.

EM-type accelerators do not actually use the EM algorithm but do use EM ideas, for example, the derivatives of the complete-data log-likelihood. Jamshidian and Jennrich categorized the quasi-Newton method described in the previous section which uses the complete-data expected information matrix $\mathcal{I}_c(\hat{\theta}^{(k)}; y)$ as the initial value B_0. Louis (17) proposed an approximate Newton method that uses the observed incomplete-data information matrix $\mathcal{I}(\hat{\theta}^{(k)}; y)$. To deal with the difficulty for computing $\mathcal{I}(\hat{\theta}^{(k)}; y)$ directly, Louis (17) gave an approximation formula (see also Ref. 1). However, it has been pointed out that Louis's approximation can also be difficult to compute in many applications (7,13).

Readers will find more acceleration methods in Refs. 1 and 13.

The EM algorithm is very attractive because of its stability and simplicity. For accelerating its slow convergence with stability and global convergence, a line search needs to be employed with any acceleration method, then, the method loses simplicity of the EM algorithm. Jamshidian and Jennrich (13) pointed out that

> For any algorithm there are two types of cost: thinking costs associated with deriving and implementing the algorithm and computer costs associated with the use of computer resources and of our patience in waiting for output. One is motivated to accelerate the EM algorithm only when the computer cost is too high. The hope is that a reduction in computing costs will be worth the increase in thinking cost associated with the accelerator. This is a personal choice.

If one makes the estimation program which would be used by many users, the basic EM algorithm would not be the choice. Then, one would at least need to compute the incomplete-data log-likelihood (the objective function) and its gradient to get faster convergence speed with stability and global convergence (cf. Refs. 7, 13, and 16), but would not necessarily need the observed incomplete-data information matrix, i.e., the Hessian matrix of the objective function, or its approximation.

REFERENCES

1. McLachlan, G. J., Krishnan, T. (1997). *The EM Algorithm and Extensions*. New York: Wiley.
2. Meng, X. L., van Dyk, D. (1997). The EM algorithm—an old folk song sung to a fast new turn. *J. R. Stat Soc. Ser. B* 44:226–233.
3. Lindstrom, M. J., Bates, D. M. (1988). Newton–Raphson and EM algorithm for linear mixed-effects models for repeated-measures data. *J. Am. Stat. Assoc.* 83:1014–1022.
4. Wright, M. H., Gill, P. E. (1997). *Practical Optimization*. London: Academic Press.
5. Nocedal, J. (1992). *Theory of Algorithms for Unconstrained Optimization. Acta Numer.* Vol. 1. London: Cambridge University Press, pp. 199–242.
6. Nocedal, J., Wright, S. J. (1999). *Numerical Optimization*. New York: Springer.
7. Minami, M. (1993). Variance Estimation for Simultaneous Response Growth Curve Models. Ph.D. Dissertation, University of California at San Diego, La Jolla, CA.
8. Saito, S., Iwasaki, H. (1995). The EM-REML iteration equations for estimating variance components in a sire and dam model. *Jpn. J. Biom.* 16:1–8.
9. Laird, N. M., Ware, J. H. (1982). Random effects models for longitudinal data. *Biometrics* 38:963–974.

10. Dempster, A. P., Laird, N. M., Rubin, D. B. (1977). Maximum likelihood from incomplete data via the EM algorithm. *J. R. Stat. Soc. Ser. B.* 39:1–22.
11. Everitt, B. S., Hand, D. J. (1981). *Finite Mixture Distributions.* London: Chapman & Hall.
12. Jordan, M. I., Xu, L. (1995). Convergence results for the EM approach to mixtures of experts architectures. *Neural. Netw.* 8:1409–1431.
13. Jamshidian, M., Jennrich, R. I. (1997). Acceleration of the EM algorithm by using quasi-Newton methods. *J. R. Stat. Soc. Ser. B.* 59:569–587.
14. Laird, N., Lange, N., Stram, D. (1987). Maximum likelihood computation with repeated measures: application of the EM algorithm. *J. Am. Stat. Assoc.* 82:97–105.
15. Lansky, D., Casella, G. (1990). *Improving the EM algorithm. Computing Science and Statistics: Proc. 22nd Symp. Interface.* New York: Springer, pp. 420–424.
16. Jamshidian, M., Jennrich, R. I. (1993). Conjugate gradient acceleration of the EM algorithm. *J. Am. Stat. Assoc.* 88:221–228.
17. Louis, T. A. (1982). Finding the observed information matrix when using the EM algorithm. *J. R. Stat. Soc. Ser. B.* 44:226–233.
18. Meilijson, I. (1989). A fast improvement to the EM algorithm on its own terms. *J. R. Stat. Soc. Ser. B* 51:127–138.

8

EM Algorithm in Neural Network Learning

Noboru Murata
Waseda University, Tokyo, Japan

Shiro Ikeda
The Institute of Statistical Mathematics, Tokyo, Japan

1 INTRODUCTION

One of the crucial purposes of neural network research is to clarify the mechanism of biological systems that can intricately adapt to changing environments, by constructing simple mathematical models. Also, as a consequence, many applications of neural networks are discussed in various engineering fields (see, for example, Ref. 1).

Its important characteristics are summarized as follows: the network is a massive parallel distributed system composed from simple homogeneous processing units, and connections between processing units are adapted by local mutual interactions of units without global information of the whole network.

Because the models proposed so far are rather simplified ones, the models are not necessarily adequate from the biological point of view, but they are preferable for mathematical treatments. It is well known that multilayered perceptrons can approximate any function with arbitrary accuracy when the number of hidden units is sufficiently large (e.g., Ref. 2); accordingly, they are frequently used for practical applications. The learning procedure based on given examples is optimizing a certain objective

95

function, and we can see a similarity to the estimating function method, namely the neural networks can be regarded as statistical models and the learning procedures can be regarded as statistical inferences. Therefore the neural networks are involved with various problems in numerous fields including statistics.

In neural networks, information are usually transmitted from input units to hidden units which do not interact directly with the outer world; then, after undergoing transformations, they arrive at output units. In the learning procedure, the states of hidden units, invisible from the outside, have to be estimated in some way, then parameter updates are carried out. These steps are closely related with the EM algorithm; in fact, the contrivance of the EM algorithm actually appears in neural network learning implicitly or explicitly.

In this article, we first review the EM algorithm from the geometrical viewpoint based on the em algorithm proposed in Ref. (7). This geometrical concept is important to interpret various learning rules in neural networks. Then, we give some examples of neural network models in which the EM algorithm implicitly appears in the learning process. From the biological point of view, it is an important problem that the EM algorithm can be appropriately implemented on real biological systems. This undertaking is usually difficult, and with a special model, the Helmholtz machine, we briefly discuss the tradeoff between statistical models and biological models. In the end, we show two models of neural networks in which the EM algorithm is adopted for learning explicitly. These models are mainly proposed for practical applications, not for biological modeling, and they are applied for complicated tasks such as controlling robots.

2 EM ALGORITHM AND em ALGORITHM

2.1 Concept of Information Geometry

Before discussing the geometrical structure of the EM algorithm, we briefly review the framework of the information geometry. The information geometry is a new approach for providing a geometrical understanding to statistical inferences and hypothesis testing—by applying the differential geometrical method to the space of probability density functions. In this section, we try to give a naive explanation of the information geometry, and the readers are referred to Refs. (3–6) for more rigorous mathematical treatment.

Let us consider the estimation problem with a parametric model. Let S be a space of the probability density function $p(x)$ of a random variable X. Strictly speaking, the distributions in S are supposed to satisfy certain regularity conditions, such as strict positivity and differentiability on their domain; in general, however, here we presume that S contains any probability density functions including empirical distributions. Let $p(x;\theta)$ be a probability density function in S parameterized by θ. A set of parametric models $\{p(x;\theta)\}$ constitutes a submanifold in the space S, and we call it a model manifold M.

From given observations $X_1,\ldots, X_T$, we can empirically calculate various statistics, and by using these statistics as a coordinate system of S, we can specify a distribution in S. Thus a set of observations corresponds to one point in the space of distributions S; however, this point is not necessarily included in the model manifold M.

In order to choose a proper parameter of the model, we have to define the closest point in the model manifold M from a point in S in some sense. This procedure is regarded as a projection from a point in the space S to a point in the model manifold M (Fig. 1).

If the space is linear and the specified model manifold is a linear subspace, the closest point can be obtained by an orthogonal projection

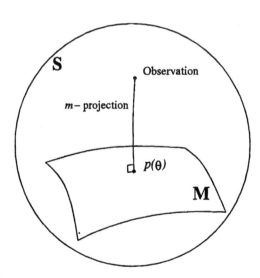

Figure 1 Geometrical interpretation of statistical inference.

from a point to the subspace. However, the space S, which we consider, is generally a "curved" space. To define a projection in the curved space, we have to extend the notion of straight lines and have to define the inner product and the orthogonality of the tangent vectors as follows.

First, let us start from the definition of "straight" lines in S. In the "curved" space, geodesics plays the role of straight lines, but the definition is not unique. When the Kullback–Leibler divergence is adopted as a statistical distance for measuring distributions, the m-geodesic and the e-geodesic play the most important roles. The m-geodesic is defined as a set of interior points between two probability density functions $p(x)$ and $q(x)$,

$$r(x; t) = (1 - t) \cdot p(x) + t \cdot q(x), \quad 0 \le t \le 1. \tag{1}$$

The e-geodesic is also defined as a set of interior points between $p(x)$ and $q(x)$, but in the sense of the logarithmic representation

$$\log r(x; t) = (1 - t)\log p(x) + t \cdot \log q(x) - \phi(t) \quad 0 \le t \le 1, \tag{2}$$

where $\phi(t)$ is the normalization term to make $r(x;t)$ a probability density function, defined by

$$\phi(t) = \log \int p(x)^{1-t} q(x)^t dx. \tag{3}$$

Similar to the definition of "straight" lines, the notion of planes can be extended as follows. Let us consider a mixture family of distributions spanned by n distinct probability density functions $p_i(x)$,

$$M_m = \left\{ p(x; \theta) = \sum_{i=1}^{n} \theta_i p_i(x), \quad \theta_i > 0, \quad \sum_{i=1}^{n} \theta_i = 1 \right\} \tag{4}$$

It is easily seen that any m-geodesic, which connects two arbitrarily chosen distributions in M_m, is included in M_m. That means the manifold is composed from "straight" lines, and M_m is a "flat" subset of S in the sense of the straightness induced by m-geodesic.

Similarly, for an exponential family such as

$$M_e = \left\{ p(x; \theta) = \exp\left(\sum_{i=1}^{n} \theta_i r_i(x) - \psi(\theta) \right) \right\}, \tag{5}$$

any e-geodesic connecting any two points in M_e is included in M_e, therefore the subset M_e is also "flat" in another sense.

The above explanation of "flatness" is just intuitive, and the m-flatness and e-flatness should be rigidly defined by using the connection in the differential geometry. For more detailed explanation, see textbooks on information geometry, for example, Refs. (3–6).

Because the concept of "flatness" is introduced in the space S as mentioned above, we next explain the orthogonal projection by defining tangent vectors and the inner product. Let a tangent vector be an infinitesimal change of the logarithmic probability density function $\log p(x)$, and let us define the inner product by the correlation of tangent vectors

$$E_p(\partial_\alpha \log p(X) \cdot \partial_\beta \log p(X)), \tag{6}$$

where ∂_α is differential along with the direction α. For example, a tangent vector along a geodesic with a parameter t is defined by

$$\begin{aligned}
\partial_t \log r(x; t) &= \frac{\partial_t r(x; t)}{r(x; t)} \\
&= \frac{\frac{d}{dt}\{(1 - t) \cdot p(x) + t \cdot q(x)\}}{r(x; t)} \\
&= \frac{q(x) - p(x)}{r(x; t)}
\end{aligned} \tag{7}$$

for m-geodesic, and

$$\begin{aligned}
\partial_t \log r(x; t) &= \frac{d}{dt}\{(1 - t) \cdot \log p(x) + t \cdot \log q(x) - \phi(t)\} \\
&= \log q(x) - \log p(x) - \frac{d}{dt}\phi(t)
\end{aligned} \tag{8}$$

for e-geodesic. The tangent vectors of the model manifold are naturally defined by the derivatives with respect to the model parameter θ

$$\begin{aligned}
\partial_{\theta_i} \log p(x; \theta) &= \frac{\partial}{\partial \theta_i} \log p(x; \theta) \\
&= \frac{\partial_{\theta_i} p(x; \theta)}{p(x; \theta)},
\end{aligned} \tag{9}$$

where θ_i is the i-th element of the parameter θ. In the following, we use the symbol ∂ to denote the differential operator without notice.

Now let us define two kinds of projections: the m-projection and the e-projection. Consider the m-geodesic from a point q in S to a point $p(\hat{\theta})$ in

M. When the m-geodesic and the model manifold M is orthogonal at the point $p(\hat{\theta})$, $p(\hat{\theta})$ is called the m-projection from q onto M. Roughly speaking, the point $p(\hat{\theta})$ in M is the closest point from q when the distance is measured by the length of the m-geodesic, and this is equivalently stated that the Kullback–Leibler divergence from q to $p(\theta)$

$$
\begin{aligned}
D(q, p(\theta)) &= \int q(x) \log \frac{q(x)}{p(x; \theta)} \, dx \\
&= E_q[\log q(X) - \log p(X; \theta)]
\end{aligned}
\tag{10}
$$

is minimized at $\hat{\theta}$. It is easily checked that at point $\hat{\theta}$, which minimizes the Kullback–Leibler divergence, all partial derivatives vanish

$$
\begin{aligned}
\partial_{\theta_i} D(q, p(\theta))\big|_{\theta=\hat{\theta}} &= -E_q\left[\partial_{\theta_i} \log p(X; \hat{\theta})\right] \\
&= 0
\end{aligned}
\tag{11}
$$

and the inner product of the tangent vector along m-geodesic at $p(\hat{\theta})$

$$
\begin{aligned}
\partial_t \log r(x; t)\big|_{t=0} &= \frac{q(x) - p(x; \hat{\theta})}{r(x; t)}\bigg|_{t=0} \\
&= \frac{q(x) - p(x; \hat{\theta})}{p(x; \hat{\theta})}
\end{aligned}
\tag{12}
$$

and the tangent vectors along the model manifold at $p(\hat{\theta})$

$$
\partial_{\theta_i} \log p(x; \theta)\big|_{\theta=\hat{\theta}} = \frac{\partial_{\theta_i} p(x; \hat{\theta})}{p(x; \hat{\theta})}
\tag{13}
$$

are calculated as

$$
\begin{aligned}
&E_{p(\hat{\theta})}[\partial_t \log r(X; 0) \cdot \partial_{\theta_i} \log p(X; \hat{\theta})] \\
&= \int \left(\frac{q(x) - p(x; \hat{\theta})}{p(x; \hat{\theta})}\right) \partial_{\theta_i} \log p(x; \hat{\theta}) p(x; \hat{\theta}) dx \\
&= E_q[\partial_{\theta_i} \log p(X; \hat{\theta})] - E_{p(\hat{\theta})}[\partial_{\theta_i} \log p(X; \hat{\theta})] \\
&= 0,
\end{aligned}
\tag{14}
$$

therefore the m-geodesic between q and $p(\hat{\theta})$ is orthogonal to the model manifold. Note that in the above calculation, we assume that differentials and integrals commute under the regularity conditions for the model, and we use the relation

$$
\begin{aligned}
E_{p(\theta)}[\partial_{\theta_i} \log p(X; \theta)] &= \partial_{\theta_i} \int p(x; \theta)\mathrm{d}x \\
&= 0.
\end{aligned}
\tag{15}
$$

As a special case, when q is the empirical distribution, the m-projection coincides with the maximum likelihood estimation (Fig. 1).

Meanwhile, the e-projection is also defined in the sense of e-geodesic, and it is equivalent to seeking the point, which minimizes

$$
D(p(\theta), q) = \int p(x; \theta)\log \frac{p(x; \theta)}{q(x)}\,\mathrm{d}x.
\tag{16}
$$

Note that the Kullback–Leibler divergence is not symmetric and the m-projection and e-projection use the Kullback–Leibler divergence in different order.

Like the m-projection, the e-geodesic and the model manifold are orthogonal at $p(x; \hat{\theta})$ in the case of the e-projection. Knowing that the tangent vector along e-geodesic is

$$
\begin{aligned}
\left.\partial_t \log r(x; t)\right|_{t=0} &= \log q(x) - \log p(x; \hat{\theta}) - \left.\frac{\mathrm{d}}{\mathrm{d}t}\phi(t)\right|_{t=0} \\
&= \log q(x) - \log p(x; \hat{\theta}) - E_{p(\hat{\theta})}[\log q(X) \\
&\quad - \log p(X; \hat{\theta})],
\end{aligned}
\tag{17}
$$

the inner product of the tangent vectors of the e-geodesic and the model manifold becomes

$$
\begin{aligned}
&E_{p(\hat{\theta})}[\partial_t \log r(X; 0) \cdot \partial_{\theta_i} \log p(X; \hat{\theta})] \\
&= \int \partial_{\theta_i} p(x; \hat{\theta})\Big\{\log q(x) - \log p(x; \hat{\theta}) \\
&\quad -E_{p(\hat{\theta})}\Big[\log q(x) - \log p(x; \hat{\theta})\Big]\Big\}\mathrm{d}x \\
&= \int \partial_{\theta_i} p(x; \hat{\theta})\Big\{\log q(x) - \log p(x; \hat{\theta})\Big\}\mathrm{d}x
\end{aligned}
\tag{18}
$$

and from the fact that

$$\partial_{\theta_i} D(p(\theta), q)\Big|_{\theta=\hat{\theta}} = \int \partial_{\theta_i} p(x; \hat{\theta}) \log p(x; \hat{\theta}) dx + \int \partial_{\theta_i} p(x; \hat{\theta}) dx$$

$$- \int \partial_{\theta_i} p(x; \hat{\theta}) \log q(x) dx \qquad (19)$$

$$= \int \partial_{\theta_i} p(x; \hat{\theta}) (\log p(x; \hat{\theta}) - \log q(x)) dx$$

$$= 0,$$

their orthogonality is confirmed.

Note that the uniqueness of the projection depends on the curvature of the model manifold. For example, when the model manifold is e-flat, the m-projection is uniquely determined. This typical example corresponds to the situation that in the Euclidean space, the orthogonal projection from a point to a plane is uniquely determined.

2.2 Geometrical Understanding of EM Algorithm

Let us consider the situation that a part of a random vector X can be observed and the rest cannot be observed. The visible variables are denoted by X_V the hidden variables are denoted by X_H, and all the vectors are written as $X = (X_V, X_H)$. The problem is to determine the parameter θ of the statistical model $p(x; \theta) = p(x_V, x_H; \theta)$ only from the observations $\{x_{V,1}, x_{V,2}, \ldots, x_{V,T})$. In this section, we consider the parameter estimation problem with hidden variables from the geometrical point of view.

When there are hidden variables which cannot be observed, it is impossible to calculate all the statistics needed to specify a point in the space S only from the observed data. In this case, we first consider the marginal distribution of the visible variables and gather all the distributions which have the same marginal distribution with the empirical distribution of the visible variables. Intuitively speaking, the set of these distributions conditioned by the marginal distribution represents observed visible data, and it is called the data manifold D (Fig. 2). Here, we introduce a new parameter η to specify the point in the data manifold D. Let $q(x_V)$ be the marginal distribution of x_V. All the points in D have the same marginal distribution and any point in D can be represented as

$$q(x_V, x_H; \eta) = q(x_V) q(x_H | x_V; \eta), \qquad (20)$$

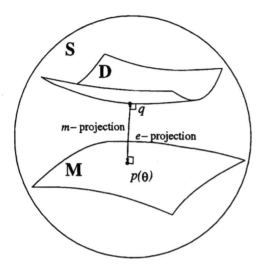

Figure 2   Observation manifold and model manifold.

thereby η can be also regarded as the parameter of the conditional probability density function $q(x_H|x_V;\eta)$.

A natural way of choosing a point in the model manifold M is adopting the closest point in M from the data manifold D. It can be achieved by measuring the statistical distance between a point $q(\eta)$ in D and a point $p(\theta)$ in M with the Kullback–Leibler divergence

$$D(q(\eta), p(\theta)) = \int q(x_V, x_H; \eta) \log \frac{q(x_V, x_H; \eta)}{p(x_V, x_H; \theta)} \, dx_V dx_H, \qquad (21)$$

and obtaining the points $\hat{\eta}$ and $\hat{\theta}$, which minimize the divergence. The em algorithm is a method of solving this estimation problem by applying e-projection and m-projection repeatedly (Fig. 3).

The procedure is composed of the following two steps.

e-step. Apply e-projection from θ_t (the t-th step estimate of θ) to D, and obtain η_{t+1} (the $t+1$-th step estimate of η).

$$\eta_{t+1} = \underset{\eta}{\mathrm{argmin}} \ D(q(\eta), p(\theta_t)). \qquad (22)$$

m-step. Apply m-projection from η_{t+1} to M, and obtain θ_{t+1} (the $t+1$-th step estimate of θ).

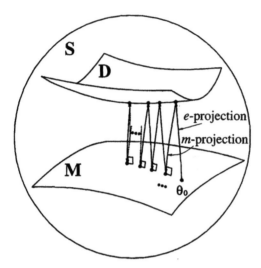

Figure 3 The em algorithm.

$$\theta_{t+1} = \underset{\theta}{\arg\min}\ D(q(\eta t + 1), p(\theta_t)). \tag{23}$$

Starting from an appropriate initial value θ_0, the procedure is expected to converge to the optimal value after sufficiently numerous iterations. If the model manifold is e-flat and the data manifold is m-flat, it is shown that in each step, the projection is uniquely determined (7); however, there are local minima in general and the algorithm can converge to the suboptimal point. The uniqueness of the solution is a subject of discussion depending on the model, such as the EM algorithm.

Note that the procedure in the e-step is equivalent to minimizing

$$
\begin{aligned}
D(q(\eta), p(\theta)) &= \int q(x_V) q(x_H|, x_V; \eta) \\
&\quad \times \log \frac{q(x_V) q(x_H|x_V; \eta)}{p(x_V; \theta_t) p(x_H|, x_V; \theta_t)}\, dx_V dx_H \\
&= \int q(x_V) \log \frac{q(x_V)}{p(x_V; \theta_t)}\, dx_V \\
&\quad + \int q(x_V) q(x_H|, x_V; \eta) \log \frac{q(x_H|x_V; \eta)}{p(x_H|x_V; \theta_t)}\, dx_V dx_H
\end{aligned}
\tag{24}
$$

$$= \int q(x_V) \log \frac{q(x_V)}{p(x_V; \theta_t)} dx_V$$

$$+ \int q(x_V) D(q(x_H|, x_V; \eta), p(x_H|, x_V; \theta_t)) dx_V,$$

and this is reduced to minimizing the conditioned Kullback–Leibler divergence $D(q(x_H|x_V;\eta), p(x_H|x_V;\theta_t))$; therefore, in usual cases, we can use

$$q(x_H|x_V, \eta_{t+1}) = p(x_H|x_V, \theta_t) \tag{25}$$

because of the positivity of the Kullback–Leibler divergence.

On the other hand, the EM algorithm consists of the Expectation step (E-step) and the Maximization step (M-step), and it gives the maximum likelihood estimate or a locally maximum point of the likelihood function by alternatively applying these two steps.

Starting from an appropriate initial value θ_0, and denoting the t-th step estimate of the parameter by θ_t, the E-step and the M-step are defined as follows.

E-step. Calculate $Q(\theta, \theta_t)$ defined by

$$Q(\theta, \theta_t) = \frac{1}{T} \sum_{k=1}^{T} \left\{ \int p(x_H|x_{V,k}; \theta_t) \log p(x_{V,k}, x_H; \theta) dx_H \right\}. \tag{26}$$

M-step. Find θ_{t+1} which maximizes $Q(\theta, \theta_t)$ with respect to θ,

$$\theta_{t+1} = \underset{\theta}{\operatorname{argmax}} \ Q(\theta, \theta_t). \tag{27}$$

The EM algorithm can be also seen as a motion on the data manifold and the model manifold (Fig. 4). In the M-step, the estimate is obtained by the m-projection from a point in the data manifold to a point in the model manifold, and this operation is equal to the m-step. In the E-step, however, the conditional expectation is calculated and this operation is slightly different from the e-projection in the e-step.

Let $q(x_V)$ be the empirical distribution of the visible variables. Suppose $q(x_H|x_V, \eta_{t+1}) = p(x_H|x_V, \theta_t)$ holds in the e-step, then the objective function evaluated in the m-step is written as

$$D(q(\eta_{t+1}), p(\theta))$$

$$= \int q(x_V) p(x_H|x_V; \theta_t) \log \frac{q(x_V) p(x_H|x_V; \theta_t)}{p(x_V, x_H; \theta)} dx_V dx_H$$

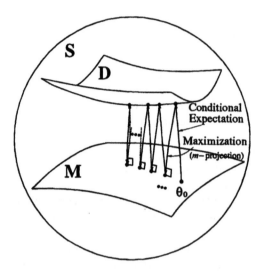

Figure 4 The EM algorithm.

$$= \int q(x_V)p(x_H|x_V; \theta_t)\log q(x_V)p(x_H|x_V; \theta_t)\mathrm{d}x_V\mathrm{d}x_H - Q(\theta, \theta_t).$$

$$(28)$$

This shows that the m-step and the M-step are equivalent if the first term can be properly integrated. Intuitively speaking, the problems occur when the integrals including the empirical distribution, which is a sum of delta functions, are not appropriately defined. In Ref. (7), the case where S is an exponential family and the model manifold is a curved exponential family embedded in S is considered, and it is shown that the E-step and the e-step give different estimates. This result mainly comes from the fact that the expectation of the hidden variables and the expectation conditioned by the visible variables do not agree

$$E_{q(\eta)}(X_H) \neq E_{q(\eta)}(X_H|x_V = E_{q(\eta)}(X_V)).$$

$$(29)$$

Actually, this example is artificial and special, and in most practical cases, the E-step and the e-step coincide.

3 EM ALGORITHM AS LEARNING RULES

The aim of neural network learning is to adapt the network parameters so that the network can reproduce well the given input–output examples. This procedure can be regarded as a statistical inference to estimate the network parameters. In most cases, neural network models have hidden units which do not interact directly with the outside world, and to which the teacher signals are not explicitly given; therefore the states of hidden units can be regarded as hidden variables. In this section, we focus on the relationship between the EM algorithm and neural networks with particular structures.

3.1 Back-Propagation Learning

The multilayered perceptron is a typical neural network model with a simple structure. In this section, we focus on the three-layered perceptron which has n input units and one output unit, as shown in Fig. 5, and try to derive the learning algorithm. Let $x \in R^n$ be an input vector, $y \in R$ an output, and $z \in R^m$ an output vector of hidden units.

The input–output relation of the neural network is denoted by $y = g(x)$, and is characterized by the relationship among the input, the output of the hidden units, and the output as

$$z_i = f(\textstyle\sum_j w_{ij} x_j)$$

$$f(u) = \frac{1}{1+e^{-u}}$$

(30)

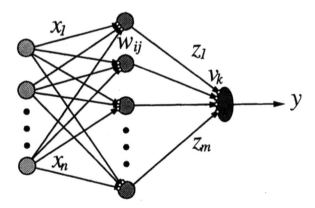

Figure 5 Perceptron.

$$g(x) = v \cdot z = \sum_j v_j z_j. \tag{31}$$

For given input–output pairs $(x_1, y_1), \ldots, (x_T, y_T)$ as examples, the purpose of learning is to find a parameter $W = \{w_{ij}\}$, $v = \{v_k\}$ with which the neural network can approximate the given examples in the best way possible. Typically, the sum of squared errors

$$E = \frac{1}{T} \sum_{k=1}^{T} (y_k - g(x_k))^2$$

is adopted as a loss function, and the parameter which minimizes the loss function is defined as optimal. In general, this is a nonlinear optimization problem and it cannot be solved arithmetically, therefore some iterative methods should be applied. A simple implementation is to use the gradient descent method, which is called the error back-propagation method. The parameter update rule is written as

$$v_{i,t+1} = v_{i,t} + \Delta v_i \tag{32}$$

$$w_{ij,t+1} = w_{ij,t} + \Delta v_i \tag{33}$$

$$\Delta v_i \propto -\frac{\partial E}{\partial v_i} = -2\frac{1}{T}\sum_{k=1}^{T}(y_k - g(x_k))z_i \tag{34}$$

$$\Delta w_{ij} \propto -\frac{\partial E}{\partial w_{ij}} = -2\frac{1}{T}\sum_{k=1}^{T}(y_k - g(x_k))v_i f'\left(\sum_{j'} w_{ij'} x_{j',k}\right) x_{j,k}. \tag{35}$$

In the update rule for the weight between the input units and the hidden units, i.e., Δw_{ij}, the term

$$(y_k - g(x_k))v_i \tag{36}$$

corresponds to the pseudoerror of the hidden unit, which is calculated as a weighted share of output error, and this is why it is called "error back-propagate." For multilayered perceptrons, which have more than three layers, the errors of the hidden units are calculated backward from the output layer to the input layer.

In the following, we rewrite the problem with statistical formulation, and derive the error back-propagation from the EM algorithm. To treat z_i

and y as random variables, we introduce additive Gaussian random variables, as follows

$$z_i = f\left(\sum_j w_{ij}x_j\right) + n_i \tag{37}$$

$$y = v \cdot z + n \tag{38}$$

$$n_1, \ldots, n_m, n \sim N(0, \sigma^2),$$

then the joint distribution of y, z, and the marginal distribution of y are written as

$$p(y, z|x; \theta) = \frac{1}{\sqrt{2\pi\sigma^2}^{m+1}} \times \exp\left\{-\frac{1}{2\sigma^2}(y - v \cdot z)^2\right.$$

$$\left. -\frac{1}{2\sigma^2}\sum_{i=1}^{m}\left(z_i - f\left(\sum_j w_{ij}x_j\right)\right)^2\right\} \tag{39}$$

$$p(y|x; \theta) = \int p(y, z|x, \theta)dz = \frac{1}{\sqrt{2\pi(1 + |v|^2)\sigma^2}}$$

$$\times \exp\left\{-\frac{1}{2(1 + |v|^2)\sigma^2}(y - g(x))^2\right\}, \tag{40}$$

where $\theta = (W, v)$.

Now, we apply the EM algorithm regarding $z_1, \ldots, z_m$ as the hidden variables. However, it is not possible to directly solve the parameter that maximizes $Q(\theta, \theta_t)$ in the M-step, we use the gradient descent to find the maxima. Or before completing the maximization of Q, we proceed to the next E-step as the Generalized EM (GEM) algorithm (8). Knowing that the conditional distribution of the hidden variables is written as

$$p(z|y, x; \theta) = \frac{p(y, z|x; \theta)}{p(y|x; \theta)}$$

$$= \frac{\sqrt{1 + |v|^2}}{(2\pi\sigma^2)^{m/2}} \exp\left(-\frac{1}{2\sigma^2}(z - r)^T(I + vv^T)(z - r)\right) \tag{41}$$

$$r = \left(\frac{I - vv^T}{1 + |v|^2}\right)(yv - f), \tag{42}$$

$$f = \left(f\left(\sum_j w_{1j}x_j \right), \ldots, f\left(\sum_j w_{mj}x_j \right) \right)^T, \tag{43}$$

where T (in superscript) denotes the transpose, the E-step and the M-step, described as follows.

E-step. Calculate the objective function

$$Q(\theta, \theta_t) = \frac{1}{T} \sum_{k=1}^{T} \left\{ \int p(z|y_k, x_k; \theta_t)(\log p(z|y_k, x_k; \theta))dz \right\}$$

$$+ \frac{1}{T} \sum_{k=1}^{T} \log p(y_k|x_k; \theta) \tag{44}$$

$$= \frac{1}{T\sigma^2} \sum_{k=1}^{T} \left\{ (r_t - r)^T(I + vv^T)(r_t - r) \right.$$

$$+ \mathrm{tr}\left(I - \frac{v_t v_t^T}{1 + |v_t|^2} \right)(I + vv^T) \right\} \tag{45}$$

$$- \frac{1}{(2\pi(1 + |v|^2)\sigma^2)^{1/2}} E - \frac{1 + m}{2} \log(2\pi\sigma^2).$$

M-step. Update the parameter along with the gradient

$$\Delta v_i \propto \frac{\partial Q(\theta, \theta_t)}{\partial v_i} = -\frac{1}{T\sigma^2} \sum_{k=1}^{T} (y_k - g(x_k))f\left(\sum_j w_{ij}x_{j,s} \right) \tag{46}$$

$$\Delta w_{ij} \propto \frac{\partial Q(\theta, \theta_t)}{\partial w_{ij}} = -\frac{1}{T\sigma^2} \sum_{k=1}^{T} (y_k - g(x_k))v_i f'$$

$$\times \left(\sum_{j'} w_{ij'}x_{j',s} \right)x_{j,s}. \tag{47}$$

This update rule coincides with the back-propagation learning rule except for constant multiplication.

3.2 Boltzmann Machine

The Boltzmann machine is a quite simple neural network designed to extract a stochastic structure of given data (9,10) (Fig. 6). The model is

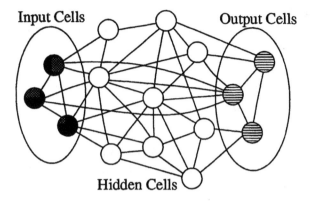

Figure 6 Boltzmann machine.

composed of n units and all the units are mutually connected. Each unit works in probabilistic way and outputs a binary value, 0 or 1. The structure of the Boltzmann machine is homogeneous and each unit works identically; however, the units are classified into three groups: the input units, the hidden units, and the output units. Signals from outside are relayed to the input units, and computational results are produced from the output units, and the hidden units do not directly interact with the external environment.

Let u_i be the internal state of the i-th unit, and let x_i be its output. The connection strength between the i-th unit and the j-th unit is denoted by w_{ij}, and we assume there is no self-connection. The internal states are calculated by the equation

$$u_i = \sum_{j \neq i} w_{ij} x_j - h_i, \tag{48}$$

where h_i is a threshold. Specifically, the internal state is determined by the weighted sum of the outputs of other units. In the following discussion, we add a special unit x_0, which always emits 1, and rewrite $w_{ii} = 0$ and $w_{i0} = h_i$; thus we use the simplified equation

$$u_i = \sum_{j=0}^{n} w_{ij} x_j \tag{49}$$

to describe the internal state. Each unit works in an asynchronous manner, which means that at each step, only one unit is updated from x_i to x_i' stochastically with the rule

$$
\begin{aligned}
p(x'_i = 1|u_i) &= \frac{1}{1 + \exp(-u_i/T)} \\
p(x'_i = 0|u_i) &= \frac{\exp(-u_i/T)}{1 + \exp(-u_i/T)},
\end{aligned}
\tag{50}
$$

where T denotes "temperature" and plays a role of controlling the rate of stochasticity. At the limit $T=0$, the units behave deterministically, which means that, depending on the sign of the internal state, 1 or 0 is emitted as an output.

The Boltzmann machine with n units can be regarded as a finite-state Markov chain with 2^n states. It is irreducible because the transition probability between any two states is not 0 from the definition, and defining the energy function as

$$
E(x) = -\sum_{i,j} w_{ij} x_i x_j,
\tag{51}
$$

where x is the output, the stationary distribution is uniquely written by the Boltzmann distribution as

$$
p(x) = \frac{1}{Z} \exp\left(-\frac{E(x)}{T}\right).
\tag{52}
$$

This is the reason why it is called "Boltzmann machine."

Let us confirm that the Boltzmann distribution cited above is actually the stationary distribution. First, assume that at the present step, the i-th unit is subject to be updated and according to Eq. (50), the output of the unit changes from x_i to x_i'. Note that the other units do not change their outputs. Because the output is subject to the Boltzmann distribution, the equation

$$
\begin{aligned}
\frac{p(x_i = 0)}{p(x_i = 1)} &= \exp\left(-\frac{E(\ldots, x_i = 0, \ldots) - E(\ldots, x_i = 1, \ldots)}{T}\right) \\
&= \exp\left(-\frac{\sum_{j \neq i} w_{ij} x_j}{T}\right) \\
&= \exp(-u_i/T)
\end{aligned}
$$

is derived, and consequently, the relation

$$p(x'_i = 1) = p(x'_i = 1, x_i = 1) + p(x'_i = 1, x_i = 0)$$

$$= p(x'_i = 1 \mid x_i = 1)p(x_i = 1) + p(x'_i = 1 \mid x_i = 0)p(x_i = 0)$$

$$= p(x'_i = 1 \mid u_i)p(x_i = 1) + p(x'_i = 1 \mid u_i)p(x_i = 0)$$

$$= p(x'_i = 1 \mid u_i)\left\{1 + \frac{p(x_i = 0)}{p(x_i = 1)}\right\}p(x_i = 1)$$

$$= \frac{1}{1 + \exp\left(-\frac{u_i}{T}\right)}\left\{1 + \exp\left(-\frac{u_i}{T}\right)\right\}p(x_i = 1)$$

$$= p(x_i = 1)$$

holds, where

$$p(x'_i = 1 \mid x_i = 1) = p(x'_i = 1 \mid x_i = 0) = p(x'_i = 1 \mid u_i) \qquad (53)$$

comes from the fact that the Boltzmann machine does not have self-connections. Therefore, it is confirmed that the Boltzmann distribution is the stationary distribution of this system.

An important feature of the Boltzmann machine is its learning mechanism. From given examples, the Boltzmann machine can extract the stochastic structure via the two-phase learning procedure. Later in this section, the states of the input units, the output units, and the hidden units are separately represented by $X = (\alpha, \beta, \gamma)$.

- **Phase I**

  Pick up an input–output pair from the given examples and clamp the input units and the output units on the chosen input–output values, then update the states of the hidden units stochastically. By repeatedly selecting the example, calculate the probability that the i-th unit and the j-th unit simultaneously output 1

  $$p_{ij} = \sum_{\alpha, \beta} q(\alpha, \beta)E(x_i x_j \mid \alpha, \beta) \qquad (54)$$

  at the equilibrium state, where $q(\alpha, \beta)$ is the empirical distribution based on the given examples.

With an appropriately small positive number ε, increase the connection weight w_{ij} by

$$\Delta w_{ij} = \varepsilon p_{ij} \tag{55}$$

according to the probability p_{ij}.

- **Phase II**

  Choose an input–output pair from the given examples and clamp only the input units on the chosen input values, then update the states of the output units and the hidden units stochastically. Calculate the probability in the same way as Phase I as

  $$p'_{ij} = \sum_{\alpha} q(\alpha)E(x_i x_j | \alpha), \tag{56}$$

  where $q(\alpha)$ is the empirical distribution of the inputs.

  Decrease the connection weight w_{ij} by

  $$\Delta w_{ij} = -\varepsilon p'_{ij} \tag{57}$$

  according to the probability p'_{ij}.

By iterating the above two phases from a certain initial connection weights, the network can obtain the weight which reproduce the same input–output distribution with the given examples. This learning procedure is derived from the gradient descent method of minimizing Kullback–Leibler divergence from the given input–output distribution to the distribution of the Boltzmann machine. Let $q(\alpha,\beta)$ be the distribution of the given examples and $p(\alpha,\beta,\gamma)$ be the distribution of the states of the Boltzmann machine. The Kullback–Leibler divergence is written as

$$
\begin{aligned}
D(w) &= \sum_{\alpha,\beta} q(\alpha,\beta)\log\frac{q(\alpha,\beta)}{p(\alpha,\beta)} \\
&= \sum_{\alpha,\beta} q(\alpha,\beta)\log\frac{q(\beta|\alpha)}{p(\beta|\alpha)}.
\end{aligned} \tag{58}
$$

Note that $q(\alpha)=p(\alpha)$ because the input units of the Boltzmann machine are clamped by the given examples. Taking into account that the stationary distribution is represented by the Boltzmann distribution, the differentials of the Kullback–Leibler divergence with respect to the weights are calculated as

$$\frac{\partial}{\partial w_{ij}} D(w) = \sum_{\alpha,\beta} \frac{q(\alpha,\beta)\partial p(\beta|\alpha)}{p(\beta|\alpha)\partial w_{ij}}$$

$$= \sum_{\alpha,\beta} \frac{q(\alpha,\beta)}{p(\beta|\alpha)} \frac{\partial}{\partial w_{ij}} \left(\sum_{\gamma} p(\beta,\gamma|\alpha) \right)$$

$$= \sum_{\alpha,\beta} \frac{q(\alpha,\beta)}{p(\beta|\alpha)} \frac{\partial}{\partial w_{ij}} \left(\sum_{\gamma} \frac{\exp(-E(\beta,\gamma|\alpha)/T)}{\sum_{\beta',\gamma'} \exp(-E(\beta',\gamma'|\alpha)/T)} \right)$$

$$= -\frac{1}{T} \sum_{\alpha,\beta} \frac{q(\alpha,\beta)}{p(\beta|\alpha)} \left\{ \sum_{\gamma} p(\beta,\gamma|\alpha)x_i x_j - p(\beta|\alpha) \sum_{\beta',\gamma'} p(\beta',\gamma'|\alpha)x_i x_j \right\}$$

$$= -\frac{1}{T} \left\{ \sum_{\alpha,\beta} q(\alpha,\beta) \sum_{\gamma} p(\gamma|\alpha,\beta)x_i x_j - \sum_{\alpha} q(\alpha) \sum_{\beta,\gamma} p(\beta,\gamma|\alpha)x_i x_j \right\}$$

$$= -\frac{1}{T} \{ p_{ij} - p'_{ij} \},$$

then by changing w_{ij} according to

$$\Delta w_{ij} \alpha - \frac{\partial}{\partial w_{ij}} D(w) \alpha p_{ij} - p'_{ij} \tag{59}$$

the Kullback–Leibler divergence decreases. In the learning procedure, Phases I and II, p_{ij} and p_{ij}' are calculated with a variation of the Monte Carlo method by making the Boltzmann machine work autonomously.

Compared with the EM algorithm, in Phase I, calculation is carried out with the conditional distribution $p(\gamma|\alpha,\beta)$, and this phase is equal to the E-step. Also, in Phase II, the gradient descent direction is calculated in the end, and the minimization of the Kullback–Leibler divergence is carried out, then this phase corresponds to the M-step. In comparison with the em algorithm, this correspondence can also be seen.

The Boltzmann machine is originally proposed as a very simple model of perceptual mechanism of human beings. For example, in natural images, there are some specific relations among the pixels in the neighborhood. Therefore even if a given image is smeared by noises, we can easily surmise the original. The Boltzmann machine is designed to capture the stochastic structure from data and because of the simple and specific structure, the learning is efficiently performed by local interactions. It is difficult to state that the Boltzmann machine is a concrete model of a brain, but it can extract the stochastic structure of the given data by repeating to memorize (Phase I) and to forget (Phase II) alternatively— thereby it is highly suggestive for understanding the mechanism of cognition and remembrance.

3.3 Helmholtz Machine

The Helmholtz machine is a kind of perceptual model which consists of the generative model and the recognition model (11–13). The concept of the Helmholtz machine is quite similar to the factor analysis model; however, because it generally employs the nonlinear regression and the learning rule is constructed only by local interaction between directly connected units, it is not always a natural model from the statistical viewpoint. The learning rule is called the Wake–Sleep algorithm, and it consists of alternative updates between the generative model and the recognition model. In the following, we explain the Wake–Sleep algorithm with a simple linear regression model as the generative and recognition models to see the mechanism (14). First, we describe the detailed model. In this case, the Helmholtz machine is equivalent to the factor analysis model.

- **Generative model**

    Let us assume the n-dimensional signal x is generated from the stochastic model

    $$x = gy + \varepsilon, \tag{60}$$

    where y is a random variable which obeys the standard normal distribution $N(0,1)$, and ε is a random noise vector subject to n-dimensional Gaussian distribution $N(0,\Sigma)$, whose mean is 0 and whose covariance matrix is a diagonal matrix $\Sigma = \mathrm{diag}(\sigma_i^2)$. The vector g corresponds to the factor loading of the factor analysis model.

- **Recognition model**

    From an observed signal x, a corresponding y is estimated by the model

    $$y = r^T x + \delta, \tag{61}$$

    where δ is a random noise subject to $N(0,s^2)$.

The purpose of the model is to reproduce the given data $\{x_1,$ $x_2,\ldots,x_N\}$ by constructing the optimal generative and recognition models. The Wake–Sleep algorithm consists of the following two phases to learn the parameters g, Σ, r, s.

- **Wake phase**

    Pick up x randomly from given examples $\{x_i\}$, and for each x generate y by using the recognition model

    $$y = r_t^T x + \delta, \qquad \delta \sim N(0, s_t^2), \tag{62}$$

    and collect a number of pairs of (x,y) by repeating this procedure. Then, update g and Σ of the generative model by

    $$g_{t+1} = g_t + \alpha \langle (x - g_t y) y \rangle \tag{63}$$

    $$\sigma_{i,t+1}^2 = \beta \sigma_{i,t}^2 + (1 - \beta) \langle (x_i - g_{i,t} y)^2 \rangle, \tag{64}$$

    where α and β are positive constants, and β is less than 1, and $\langle \cdot \rangle$ denotes the expectation with respect to collected data x and y.

- **Sleep phase**

    Generate y subject to the standard normal distribution, and produce a pseudodata x by the generative model

    $$x = g_t y + \varepsilon, \qquad \varepsilon \sim N(0, \Sigma_t). \tag{65}$$

    Collect a number of pairs of (x,y), and update the parameter r, s^2 of the recognition model by

    $$r_{t+1} = r_t + \alpha' \langle x(y - r_t^T x) \rangle \tag{66}$$

    $$s_{t+1}^2 = \beta' s_t^2 + (1 - \beta') \langle (y - r_t^T x)^2 \rangle, \tag{67}$$

    where $\langle \cdot \rangle$ denotes the expectation.

An important feature of this learning rule is that only local information is used. For example, g_i, the i-th element of g, connects y and x_i, the i-th element of x, and its update is calculated only from the values at both ends of the weight as

$$g_{i,t+1} = g_{i,t} + \alpha \langle (x_i - g_{i,t} y) y \rangle. \tag{68}$$

This restriction on locality is not necessary for a computational model; however, it is sometimes required as an information processing model of biological systems, because in real biological systems, connections between information processing units, i.e., neurons, are limited in narrow region, hence the global information cannot be used and only the local information is available.

From the geometrical point of view, the Wake–Sleep algorithm can be seen as reducing two different Kullback–Leibler divergences alternatively. In the generative model, the joint distribution of x and y is written as

$$p(y, x; \theta) = \exp\left(-\frac{1}{2}(y \quad x^T)A\begin{pmatrix} y \\ x \end{pmatrix} - \psi(\theta)\right)$$

$$A = \left(\begin{array}{c|c} 1 + g^T\Sigma^{-1}g & -g^T\Sigma^{-1} \\ \hline -\Sigma^{-1}g & \Sigma^{-1} \end{array}\right) \tag{69}$$

$$\psi(\theta) = \frac{1}{2}\left(\sum \log \sigma_i^2 + (n+1)\log 2\pi\right),$$

where $\theta = (g, \Sigma)$. Also, in the recognition model, the conditional distribution of y conditioned by x is the normal distribution $N(r^T x, s^2)$, and the distribution of x is the normal distribution $N(0, C)$ where C is calculated from the given data $x_1, \ldots, x_N$ by

$$C = \frac{1}{N}\sum_{s=1}^{N} x_s x_s^T, \tag{70}$$

therefore the joint distribution of x and y is written as

$$q(y, x; \eta) = \exp\left(-\frac{1}{2}(y \quad x^T)B\begin{pmatrix} y \\ x \end{pmatrix} - \psi(\eta)\right)$$

$$B = \frac{1}{s^2}\left(\begin{array}{c|c} 1 & -r^T \\ \hline -r & s^2 C^{-1} + rr^T \end{array}\right) \tag{71}$$

$$\psi(\eta) = \frac{1}{2}(\log s^2 + \log|C| + (n+1)\log 2\pi),$$

when $\eta = (r, s^2)$. For these two joint distributions, the Wake phase works as decreasing

$$D(q(\eta), p(\theta)) = E_{q(\eta)}\left(\log \frac{q(y, x; \eta)}{p(y, x; \theta)}\right) \tag{72}$$

with respect to θ, and the Sleep-phase works as decreasing

$$D(p(\theta), q(\eta)) = E_{p(\theta)}\left(\log \frac{p(y, x; \theta)}{q(y, x; \eta)}\right) \tag{73}$$

with respect to η. Hence both phases correspond to the m-projection.

In the case of the linear model above, although the dynamics of learning is different from the EM algorithm and the em algorithm as shown, the parameter (θ, η) converges to the maximum likelihood estimate. However, for the model, include nonlinear transformations such as

$$x = f(gy) + \varepsilon \tag{74}$$
$$y = h(r^T x) + \delta, \tag{75}$$

the Wake–Sleep algorithm does not generally converge to the maximum likelihood estimate.

The structure of the Helmholtz machine is more complex than that of the Boltzmann machine, but as a cognitive model, the Helmholtz machine is attractive because of its clearly separated structure. Mathematically, some problems remain, such as stability of learning and convergence.

4 MODULE MODELS FOR EM ALGORITHM

In the previous section, we treated models which have hidden units, and the states of hidden units are regarded as hidden variables. In this section, we review some models in which the EM algorithm is explicitly used for learning.

4.1 Mixtures of Experts

The mixture of experts is a hierarchical neural network model (15,16). It consists of some modules and the modules cooperate and compete with each other to process the information.

The model is constructed out of two parts: the expert networks and the gating networks. The gating networks can be arranged hierarchically (hierarchical mixtures of experts), but in the following, we consider the simplest case, as shown in Fig. 7.

- **Expert network**

  Each network receives the input $x \in R^m$ and generates the output $\mu_i \in R^d$.

  $$\mu_i = f_i(x; \theta_i), \quad i = 1, \ldots, K, \tag{76}$$

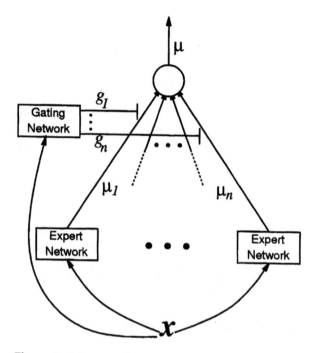

Figure 7   Mixtures of experts.

where θ_i is the parameter to be learned. Although we do not specify the structure of the network here, linear regression models and multilayered perceptrons are often used in practical cases. The parameter θ_i corresponds to the regression coefficient in the former case, and the connection weights in the latter case. We assume that the output includes an additive Gaussian noise, and the output of the network represents the mean value, namely,

$$y = \mu_i + n \tag{77}$$

where n is subject to a normal distribution. Then, the conditional distribution of y is written as

$$p(y|x, \theta_i) = \frac{1}{\sqrt{(2\pi)^d |\Sigma_i|}} \exp\left(-\frac{1}{2}(y - f_i(x; \theta_i))^T \right.$$
$$\left. \times \Sigma_i^{-1}(y - f_i(x; \theta_i))\right). \tag{78}$$

- **Gating network**

  Using an appropriate function, $s_i(x;\theta_0)$, a competitive system is constructed by

  $$g_i(x;\theta_0) = \frac{\exp(s_i(x;\theta_0))}{\sum_{j=1}^{K} \exp(s_j(x;\theta_0))}. \tag{79}$$

  Frequently, multilayered perceptrons are adopted as s_i. Knowing that g_i is positive and the sum of g_i's is 1, (g_i) can be regarded as a probability vector. The output of the whole network is given by an output of an expert network selected by the probability (g_i), or the weighted average of all the expert networks.

For given input–output pairs $\{(x_1,y_1),\ldots,(x_T,y_T)\}$, the learning rule is derived from the EM algorithm where g_i's are treated as hidden variables.

In the following, we describe the parameter update rule for linear regression experts, which are important for practical applications (17).

- **E-step**

For each input-output example (x_k,y_k), calculate the conditional distribution

$$p(i|x,y) = \frac{g_i(x,\theta_{0,t})p(y|x,\theta_{i,t})}{\sum_{j=1}^{n} g_j(x,\theta_{0,t})p(y|x,\theta_{j,t})}. \tag{80}$$

- **M-step**

Update the parameters by

$$\Sigma_{i,t+1} = \frac{\sum_{k=1}^{T} p(i|x_k,y_k)(y_k - f_i(x_k;\theta_{i,t}))(y_k - f_i(x_k;\theta_{i,t}))^T}{\sum_{k=1}^{T} p(i|x_k,y_k)} \tag{81}$$

$$\theta_{i,t+1} = R_{i,t}^{-1} e_{i,t} \tag{82}$$

$$e_{i,t} = \sum_{k=1}^{T} p(i|x_k,y_k)x_k\Sigma_{i,t}^{-1}y_k$$

$$R_{i,t} = \sum_{k=1}^{T} p(i|x_k,y_k)x_k\Sigma_{i,t}^{-1}x_k^T$$

$$X_k^T = \begin{pmatrix} x_k^T & 0 & \cdots & 0 & 1 & 0 & \cdots & 0 \\ \vdots & \ddots & & \vdots & \vdots & \ddots & & \vdots \\ 0 & 0 & \cdots & x_k^T & 0 & 0 & \cdots & 1 \end{pmatrix}$$

$$\theta_{0,t+1} = \theta_{0,t} + \delta R_{0,t}^{-1} e_{0,t} \tag{83}$$

$$e_{0,t} = \sum_{k=1}^{T} \sum_{i=1}^{K} (p(i|x_k, y_k) - g_i(x_k; \theta_{0,t})) \frac{\partial s_i(x_k; \theta_{0,t})}{\partial \theta_0}$$

$$R_{0,t} = \sum_{k=1}^{T} \sum_{i=1}^{K} g_i(x_k; \theta_{0,t})$$

$$- (1 - g_i(x_k; \theta_{0,t})) \frac{\partial s_i(x_k; \theta_{0,t})}{\partial \theta_0} \frac{\partial s_i(x_k; \theta_{0,t})^T}{\partial \theta_0}.$$

For mixtures of experts, many variations of update rules are proposed for on-line learning, or improving speed of convergence and stability.

4.2 Normalized Gaussian Network

The normalized Gaussian network (18) is composed of the first-order spline functions, which map m-dimensional inputs to d-dimensional outputs, and the radial basis function network. Sensory neurons in biological systems show strong reaction for specific stimuli from outside, and show no reaction for other signals. This domain of sensitive signals is called the receptive field and the radial basis function network simulates the receptive field in quite simple manner.

The structure of the normalized Gaussian network is described as

$$y = \sum_{i=1}^{M} (W_i x + b_i) N_i(x)$$

$$N_i(x) = \frac{G_i(x)}{\sum_{j=1}^{M} G_j(x)} \tag{84}$$

$$G_i(x) = \frac{1}{\sqrt{(2\pi)^m |\Sigma_i|}} \exp\left\{ -\frac{1}{2} (x - \mu_i)^T \Sigma_i^{-1} (x - \mu_i) \right\},$$

where N_i's are radial basis functions, and M is the number of units in the network.

Let us assume that for an input–output pair (x,y), one unit i is chosen, and let us represent this by (x,y,i). We do not know which unit is chosen, so the indicator i is regarded as a hidden variable. We assume that the probability of complete data (x,y,i) is written by

$$p(x,y,i;\theta) = \frac{1}{M\sqrt{(2\pi)^{(m+d)}\sigma_i^{2d}|\Sigma_i|}}\exp\left\{-\frac{1}{2}(x-\mu_i)^T\Sigma_i^{-1}(x-\mu_i)\right\}$$

$$\times \exp\left\{-\frac{1}{2\sigma_i^2}(y-W_ix-b_i)^2\right\}, \tag{85}$$

where $\theta = \{\mu_i,\Sigma_i,\sigma_i^2,W_i,b_i; i=1,\ldots,M\}$, and when the unit i is chosen, the mean value is given by W_ix+b_i and the variance of the output y is σ_i^2. And the conditional probability is calculated by

$$p(y|x,i;\theta) = \frac{1}{\sqrt{2\pi\sigma_i^2}^d}\exp\left\{\frac{1}{2\sigma_i^2}(y-W_ix-b_i)^2\right\} \tag{86}$$

$$p(y|x;\theta) = \sum_{i=1}^{M} N_i(x)p(y|x,i,\theta). \tag{87}$$

From this, when the input is x, the expectation of output y is given by

$$E(y|x) = \int yp(y|x;\theta)dy = \sum_{i=1}^{M}(W_ix+b_i)N_i(x) \tag{88}$$

and this is equal to the output of the network. Therefore the network is regarded as calculating the expectation of the output.

Regarding the normalized Gaussian network as a stochastic model, for given input–output examples, the maximum likelihood estimate of the network parameter can be obtained by an iterative method based on the EM algorithm (19).

- **E-step**

    Calculate the posterior probability so that the unit i is chosen under the parameter θ_t and the given data (x,y) by

    $$p(i|x,y;\theta_t) = \frac{p(x,y,i;\theta_t)}{\sum_{j=1}^{M}p(x,y,j;\theta_t)}. \tag{89}$$

- **M-step**

  Define the expectation of the log likelihood for the complete date by

  $$Q(\theta, \theta_t) = \sum_{k=1}^{T} \sum_{i=1}^{M} p(i \mid x_k, y_k; \theta_t) \log p(i \mid x_k; y_k; \theta) \qquad (90)$$

  and maximize this with respect to θ. By using the weighted average subject to the posterior distribution defined by

  $$E_i(f(x, y)) = \frac{1}{T} \sum_{k=1}^{T} f(x_k, y_k) p(i \mid x_k, y_k; \theta_t), \qquad (91)$$

  the solution is given by

  $$\mu_{i,t+1} = \frac{E_i(x)}{E_i(1)} \qquad (92)$$

  $$\sum_{i,t+1} = \frac{E_i((x - \mu_{i,t})(x - \mu_{i,t})^T)}{E_i(1)} \qquad (93)$$

  $$\tilde{W}_{i,t+1} = E_i(y\tilde{x}^T) E_i(\tilde{x}\tilde{x}^T)^{-1} \qquad (94)$$

  $$\sigma_{i,t+1}^2 = \frac{1}{d} \frac{E_i(|y - \tilde{W}_{i,t}\tilde{x}|^2)}{E_i(1)}, \qquad (95)$$

  where

  $$\tilde{W}_i = (W_i, b_i), \quad \tilde{x}^T = (x^T, 1).$$

  In practical applications, on-line learning is often employed, that is, updating parameters and observing new data are alternatively carried out. In such cases, the learning system is supposed to follow the changing environment, and E_i is replaced by a running average, i.e., a low-pass filter. Practical applications such as controlling robots are found in Ref. (20), for example.

REFERENCES

1. Rumelhart, D., McClelland, J. L., The PDP Research Group (1986). *Parallel Distributed Processing: Explorations in the Microstructure of Cognition.* Cambridge, MA: The MIT Press.

2. Cybenko, G. (1989). Approximation by superpositions of a sigmoid function. *Mathematics of Control, Signals and Systems* 2:303–314.
3. Amari, S. (1985). *Differential–Geometrical Methods in Statistics*. Vol. 28 of Lecture Notes in Statistics. Berlin: Springer-Verlag.
4. Barndorff-Nielsen, O. (1988). *Parametric Statistical Models and Likelihood, volume 50 of Lecture Notes in Statistics*. New York: Springer-Verlag.
5. Murray, M. K., Rice, J. W. (1993). *Differential Geometry and Statistics*. London; New York: Chapman & Hall.
6. Amari, S., Nagaoka, H. (2000). *Methods of Information Geometry. Translations of Mathematical Monographs*. Providence, RI: AMS, Oxford University Press.
7. Amari, S. (1995). Information geometry of the EM and em algorithms for neural networks. *Neural Networks* 8(9):1379–1408.
8. McLachlan, G. J., Krishnan, T. (1997). *The EM Algorithm and Extensions. Wiley series in probability and statistics*. New York: John Wiley & Sons, Inc.
9. Hinton, G. E., Sejnowski, T. J. (1983). Optimal perceptual inference. *Proceedings of the IEEE Conference on Computer Vision and Pattern Recognition*, pp. 448–453.
10. Ackley, D. H., Hinton, G. E., Sejnowski, T. J. (1985). A learning algorithm for Bopltzmann machines. *Cognitive Science* 9:147–169.
11. Dayan, P., Hinton, G. E., Neal, R. M. (1995). The Helmholtz machine. *Neural Computation* 7(5):889–904.
12. Hinton, G. E., Dayan, P., Frey, B. J., Neal, R. M. (1995). The "sake–sleep" algorithm for unsupervised neural networks. *Science* 268:1158–1160.
13. Neal, R. M., Dayan, P. (November 1997). Factor analysis using delta-rule wake–sleep learning. *Neural Computation* 9(8):1781–1803.
14. Ikeda, S., Amari, S., Nakahara, H. (1999). Convergence of the Wake–Sleep algorithm. In: Kearns, M. S., Solla, S. A., Cohn, D. A., eds., *Advances in Neural Information Processing Systems* 11. Cambridge, MA: The MIT Press, pp. 239–245.
15. Jacobs, R. A., Jordan, M. I., Nowlan, S. J., Hinton, G. E. (1991). Adaptive mixtures of local experts. *Neural Computation* 379–87.
16. Jordan, M. I., Jacobs, R. A. (1994). Hierarchical mixtures of experts and the EM algorithm. *Neural Computation* 6:181–214.
17. Jordan, M. I., Xu, L. (1995). Convergence results for the EM approach to mixtures of experts architectures. *Neural Networks* 8(9):1409–1431.
18. Moody, J., Darken, C. (1989). Fast learning in networks of locally-tuned processing units. *Neural Computation* 1:289–303.
19. Sato, M., Ishii, S. (2000). On-line EM algorithm for the normalized Gaussian network. *Neural Computation* 12:407–432.
20. Doya, K. (1997). Efficient nonlinear control with actor–tutor architecture. Touretzky, D. S., Mozer, M. C., Hasselmo, M. E., eds., *Advances in Neural Information Processing Systems* Vol. 9. Cambridge, MA: The MIT Press, pp. 1012–1018.

9
Markov Chain Monte Carlo

Masahiro Kuroda
Okayama University of Science, Okayama, Japan

1 INTRODUCTION

Markov chain Monte Carlo (MCMC) is the generic name of Monte Carlo integration using Markov chains and is used in Bayesian inference and likelihood inference such as evaluating the posterior distributions or the likelihood functions of interesting parameters. MCMC can be roughly classified by two methods: one is the Metropolis–Hastings (M–H) algorithm and another is the Gibbs Sampling (GS) algorithm. The Gibbs Sampling algorithm can be also regarded as a particular case of the Metropolis–Hastings algorithm.

The Metropolis–Hastings algorithm was developed by Metropolis et al. (1953) to study equilibrium properties of chemical substances and was extended to use statistical computations by Hastings (1970). The Gibbs Sampling algorithm was firstly presented by Geman and Geman (1984) in the analysis of image processing models. In order to apply to Bayesian inference for missing data, Tanner and Wong (1987) proposed the Data Augmentation (DA) algorithm, which is a special version of the Gibbs Sampling algorithm. In this paper, they demonstrated that the Bayesian computation, to be infeasible analytically, can be performed by using the iterative Monte Carlo method. Gelfand and Smith (1990) reviewed the Data Augmentation and the Gibbs Sampling algorithms and revealed that their algorithms can be found in the posterior distributions for complex statistical models and were widely available for the missing data analysis.

These MCMC methods have become standard Bayesian computational methods and increasingly popular statistical techniques of the likelihood inference dealing with missing data.

This chapter is organized as follows. In Sec. 2, we introduce important properties with Markov chains that have stationary distributions. In Secs. 3 Secs. 4 Secs. 5, we deal with three MCMC methods: the Data Augmentation, the Gibbs Sampling, and the Metropolis–Hastings algorithms. In each of these sections, we show the process to deduce the computational algorithm and illustrate simple examples to help the readers understand. Sec. 6 refers to several books and papers for the further study of MCMC methods.

2 MARKOV CHAINS

In this section, we will briefly summarize some properties with Markov chains used in MCMC methods. First, we consider Markov chains with discrete and finite state space $S = \{0, 1, 2, \ldots\}$ and, second, we deal with continuous state space.

A *Markov chain* is a stochastic process $\{\theta^{(t)} | t \geq 0\}$, which implies that the future state is independent of the past states given the current state. Then, by the Markov property, the probability of moving from $\theta^{(t)} = \phi$ to $\theta^{(t+1)} = \theta$ can be expressed by:

$$\Pr(\theta^{(t+1)} = \theta | \theta^{(t)} = \phi)$$

$$= \Pr(\theta^{(t+1)} = \theta | \theta^{(t)} = \phi, \theta^{(t-1)} = \phi_{t-1}, \ldots, \theta^{(0)} = \phi_0) \quad (1)$$

for all states ϕ, $\theta \in S$. When the Markov chain is *time-homogeneous* such that Eq. (1) does not depend on t, we can define the *transition kernel* $K(\phi, \theta)$ by the conditional probability:

$$K(\phi, \theta) = \Pr(\theta^{(t+1)} = \theta | \theta^{(t)} = \phi), \quad (2)$$

where $K(\phi, \theta)$ is non-negative and satisfies $\sum_{\theta} K(\phi, \theta) = 1$. Our attention is restricted to time-homogeneous Markov chains. Denoting the transition kernel over $(u + t)$ steps as:

$$K^{u+t}(\phi, \theta) = \Pr(\theta^{(u+t)} = \theta | \theta^{(t)} = \phi),$$

it can be obtained by:

$$K^{u+t}(\phi, \theta) = \sum_{\theta' \in S} K^u(\theta', \theta) K^t(\phi, \theta'). \tag{3}$$

Then Eq. (3) is called *Chapman–Kolmogorov equation*.

In the use of MCMC methods, the fundamental and important problem is that the distribution of $\theta^{(t)}$ converges to a stationary distribution after a finite number of iterations. For the distribution of $\theta^{(t)}$ to converge to a stationary distribution, the chain must satisfy three properties: *irreducible, aperiodic,* and *positive recurrent*. We will give the outlines of these properties. When a Markov chain is *irreducible*, the chain starting from a state ϕ can reach a state θ with a positive probability by finite steps, $K^t(\phi,\theta) > 0$ for some arbitrary step $t > 0$. For an *aperiodic* Markov chain, the chain visits a state θ without a regular period. The recurrence means that the average number of visits to a state θ is infinite, that is, $\sum_{t=0}^{\infty} K^t(\theta, \theta) = \infty$. If S is finite, the irreducible chain is recurrent. In addition, the Markov chain is called the *positive recurrent* if the expected return time to a state θ is finite. When a Markov chain is irreducible, aperiodic, and positive recurrent, the chain is called *ergodic*. For the ergodic Markov chain, there exists a probability distribution $\pi(\theta)$ such as:

$$\sum_{\phi \in S} \pi(\phi) K(\phi, \theta) = \pi(\theta) \tag{4}$$

for all states $\theta \in S$. Then $\pi(\theta)$ is said to be the *stationary distribution* of the Markov chain $\{\theta^{(t)} | t \geq 0\}$. The ergodic Markov chain has also the *limiting distribution* that satisfies:

$$\pi(\theta) = \lim_{t \to \infty} K^t(\phi, \theta) \tag{5}$$

for all states $\phi, \theta \in S$. In order to generate samples from $\pi(\theta)$, MCMC methods find the transition kernel $K(\phi,\theta)$, which satisfies Eqs. (4) and (5). Moreover, if the chain satisfies *time reversibility* such that:

$$\pi(\phi) K(\phi, \theta) = \pi(\theta) K(\theta, \phi),$$

for all states $\phi, \theta \in S$, then it holds:

$$\sum_{\phi \in S} \pi(\phi) K(\phi, \theta) = \sum_{\phi \in S} \pi(\theta) K(\theta, \phi) = \pi(\theta).$$

Consequently, the *time reversibility* provides the sufficient condition to converge to $\pi(\theta)$. In the application of MCMC methods, it is important to specify a transition kernel with the reversibility that guarantees the convergence of Markov chains.

For the ergodic Markov chain, the *law of large numbers theorem* holds and then guarantees that the ergodic average of a real-valued function $h(\theta)$:

$$\bar{h}(\theta) = \frac{1}{T}\sum_{t=1}^{T}h(\theta^{(t)}),$$

converges to the expectation $E[h(\theta)]$. This result states that even if the sequence of the chain is not independent, the ergodic average of the chain values gives strong, consistent estimates of parameters of $\pi(\theta)$.

Next, consider Markov chains with continuous state space. Then the *transition density* $K(\phi,\theta)$ moving from $\theta^{(t)} = \phi$ to $\theta^{(t+1)} = \theta$ is given by a conditional density, instead of the conditional probability (Eq. (2)), and $\int K(\phi,\theta)d\theta = 1$. For any subset $A \subset S$, the *transition kernel* $K(\phi,A)$ can be defined by:

$$K(\phi, A) = \int_{A} K(\phi, \theta)d\theta = \Pr(\theta^{(t+1)} \in A | \theta^{(t)} = \phi).$$

The stationary distribution $\pi(\theta)$ of the chain with the transition density $K(\phi,\theta)$ must satisfy:

$$\pi(\theta) = \int K(\phi, \theta)\pi(\phi)d\phi$$

for all states $\phi \in S$. For Markov chains with continuous state space, the ergodic chains must have irreducible, aperiodic, and *Harris recurrent* properties instead of positive recurrent (see Tierney, 1994). Then all of the convergence results of Markov chains with discrete state space are valid in Markov chains with continuous state space. For a more technical or theoretical discussion of Markov chains required in MCMC methods, see Robert and Casella (1999), Gamerman (1997), and Tierney (1995).

3 THE DATA AUGMENTATION ALGORITHM

The Data Augmentation algorithm is applied to Bayesian inference with missing data. The DA algorithm is very suitable when the incomplete-data posterior distribution has a complicated density but the posterior distribution given the augmented data filled in missing values has a good density to produce samples.

The basic idea of the DA algorithm is to augment the observed data by imputing the missing values and to find the posterior distribution in the framework of complete data case. We write the augmented data $x = (x_{obs}, x_{mis})$, where x_{obs} denotes observed data and x_{mis} denotes missing data. In order to obtain the posterior distribution $\pi(\theta|x_{obs})$, the DA algorithm generates a large number of imputed values from the predictive distribution $f(x_{mis}|x_{obs})$ and updates the current posterior distribution $\pi(\theta|x_{obs}, x_{mis})$ given x by the Monte Carlo method. Then $f(x_{mis}|x_{obs})$ depends on $\pi(\theta|x_{obs})$, so that it requires to derive cyclically these distributions.

The DA algorithm is the successive substitution scheme to estimate the posterior distribution and predictive distribution. The posterior distribution of θ is given by:

$$\pi(\theta|x_{obs}) = \int \pi(\theta|x_{mis}, x_{obs}) f(x_{mis}|x_{obs}) dx_{mis}, \tag{6}$$

where $\pi(\theta|x_{mis}, x_{obs})$ denotes the posterior distribution of θ given $x = (x_{obs}, x_{mis})$, and the predictive distribution of x_{mis} is obtained by:

$$f(x_{mis}|x_{obs}) = \int f(x_{mis}|x_{obs}, \phi) \pi(\phi|x_{obs}) d\phi, \tag{7}$$

where $f(x_{mis}|x_{obs}, \phi)$ is the predictive distribution conditionally on ϕ. Now substituting Eq. (7) into Eq. (6) and interchanging the integral order, we can obtain the following integral equation:

$$\pi(\theta|x_{obs}) = \int K(\phi, \theta) \pi(\phi|x_{obs}) d\phi, \tag{8}$$

where

$$K(\phi, \theta) = \int \pi(\theta|x_{obs}, x_{mis}) f(x_{mis}|x_{obs}, \phi) dx_{mis},$$

and $K(\phi, \theta)$ is the *transition kernel*. Starting the initial distribution as $\pi^{(0)}(\theta|x_{obs})$, the posterior distribution at the t-step can be obtained by the substitutive calculation:

$$\pi^{(1)}(\theta|x_{obs}) = \int K(\phi, \theta) \pi^{(0)}(\phi|x_{obs}) d\phi,$$

$$\pi^{(2)}(\theta|x_{obs}) = \int K(\phi, \theta)\pi^{(1)}(\phi|x_{obs})d\phi,$$

$$\vdots \tag{9}$$

$$\pi^{(t)}(\theta|x_{obs}) = \int K(\phi, \theta)\pi^{(t-1)}(\phi|x_{obs})d\phi.$$

From the above series of equations, the t-step posterior distribution $\pi^{(t)}(\theta|x_{obs})$ can be calculated by using the successive substitution of the $(t-1)$-step posterior distribution $\pi^{(t-1)}(\theta|x_{obs})$. When $\pi^{(t)}(\theta|x_{obs})$ reaches a stationary distribution that satisfies Eq. (8), we can find the true posterior distribution $\pi(\theta|x_{obs})$. Because the integration is infeasible analytically, the Monte Carlo approximation can be applied to the estimation of $\pi(\theta|x_{obs})$. In order to update the approximate posterior distribution, the DA algorithm carries out the following iterative scheme:

Initialization: Set the initial distribution $\pi^{(0)}(\theta|x_{obs})$.

Imputation step: Repeat the following steps for $l = 1, \ldots, L$ to obtain the imputed values of x_{mis} from the predictive distribution $f(x_{mis}|x_{obs})$.

1. Generate θ^* from the current approximated posterior distribution $\pi^{(t-1)}(\theta|x_{obs})$.
2. Generate the imputed value $x_{mis}^{(l)}$ from the conditional predictive distribution $f(x_{mis}|x_{obs}, \theta^*)$, where θ^* is obtained by the above step.

Posterior step: Update the current approximation $\pi^{(t-1)}(\theta|x_{obs})$, given $x_{mis}^{(l)}$ for $l = 1, \ldots, L$, by the Monte Carlo method:

$$\pi^{(t)}(\theta|x_{obs}) = \frac{1}{L}\sum_{l=1}^{L}\pi(\theta|x_{obs}, x_{mis}^{(l)}).$$

Until the approximated posterior distribution $\pi^{(t)}(\theta|x_{obs})$ converges to a stationary distribution $\pi(\theta|x_{obs})$, the *Imputation step* and the *Posterior step* are iterated. The *Imputation step* performs the calculation of $K(\phi,\theta)$ using sample-based approximation, and the *Posterior step* does the integration to obtain $\pi(\theta|x_{obs})$ by the Monte Carlo method. Seeing the iteration between the *Imputation step* and the *Posterior step*, we notice that the DA algorithm is the iterative simulation version of the EM algorithm: the former step corresponds to the *Expectation step* and the latter step corresponds to the *Maximization step*.

With respect to convergence of the DA algorithm, Tanner and Wong (1987) gave the following results under mild regularity conditions:

Result 1 (*Uniqueness*): The true posterior distribution $\pi(\theta|x_{\text{obs}})$ is the unique solution of Eq. (9).

Result 2 (*Convergence*): For almost any $\pi^{(0)}(\theta|x_{\text{obs}})$, the sequence of $\pi^{(t)}(\theta|x_{\text{obs}})$, for $t = 1, 2, \ldots$, obtained by Eq. (9), converges monotonically in L_1 to $\pi(\theta|x_{\text{obs}})$.

Result 3 (*Rate*): $\int |\pi^{(t)}(\theta|x_{\text{obs}}) - \pi(\theta|x_{\text{obs}})| d\theta \to 1$ geometrically in t.

Example: Genetic Linkage Model

Tanner and Wong (1987) applied the DA algorithm to a contingency table wherein two cells are grouped into one cell. Let 197 animals be categorized into four cells:

$$x = (x_1, x_2, x_3, x_4) = (125, 18, 20, 34)$$

with probabilities

$$\theta = \left(\frac{1}{2} + \frac{1}{4}p, \frac{1}{4}(1-p), \frac{1}{4}(1-p), \frac{1}{4}p\right).$$

We suppose that x has the multinomial distribution with θ. Then the likelihood of x can be expressed by a less intractable functional form:

$$f(x|\theta) \propto (2+p)^{x_1}(1-p)^{x_2+x_3}p^{x_4}.$$

By splitting the first cell x_1 into two cells y_1 and y_2 with probabilities $1/2$ and $p/4$, we recategorize x such that:

$$y = (y_1, y_2, y_3, y_4, y_5) = (x_1 - y_2, y_2, x_2, x_3, x_4).$$

Then the likelihood of y can be simplified as:

$$f(y|\theta) \propto p^{y_2+y_5}(1-p)^{y_3+y_4}.$$

We assume that the prior distribution of p has the beta distribution with hyperparameters $\alpha = (\alpha_1, \alpha_2)$. The predictive distribution of y_2 has the conditional binomial distribution with the parameter $p/(2+p)$. Then each of the distributions has the density:

$$\pi(p|\alpha) \propto p^{\alpha_1-1}(1-p)^{\alpha_2-1},$$

$$f(y_2|x_1, p) \propto \left(\frac{p}{2+p}\right)^{y_2}\left(\frac{2}{2+p}\right)^{x_1-y_2}.$$

For this multinomial model, the posterior distribution of p given y can be found by the DA algorithm:

Initialization: Setting the initial distribution $\pi^{(0)}(p|x) = \pi(p|\alpha)$.
Imputation step: Repeating the following steps for $l = 1, \ldots, L$.

1. Generate p^* from $\pi^{(t-1)}(p|x)$.
2. Generate $y_2^{(l)}$ from $f(y_2|x_1, p^*)$ and obtain $y^{(l)} = (x_1 - y_2^{(l)}, y_2^{(l)}, y_3, y_4, y_5)$.

Posterior step: Update the current approximation $\pi^{(t-1)}(p|x)$ given $y^{(l)}$ using the Monte Carlo method:

$$\pi^{(t)}(p|x) = \frac{1}{L} \sum_{l=1}^{L} \pi(p|y^{(l)}),$$

where

$$\pi(p|y^{(l)}) \propto p^{\alpha_1 + y_2^{(l)} + y_5 - 1}(1 - p)^{\alpha_2 + y_3 + y_4 - 1}.$$

The *Imputation step* and the *Posterior step* are iterated until the convergence of the DA algorithm. As the prior distribution of p, we select a flat prior distribution. From the simulated samples 2000 after a burn-in sample 400, we can obtain the posterior mean $E[p|x] = 0.6230$ as the estimate of p, and also the posterior variance $\text{Var}[p|x] = 0.0025$. Applying the EM algorithm, the MLE of p is 0.6268 and the variance is 0.0025 using the Louis method in Louis (1982).

Example: Highway Safety Research

Kuroda and Geng (2002) applied the DA algorithm to the double sampling data from Hochberg (1977). The data were the highway safety research data relating seat belt usages to driver injuries. The main sample consists of 80,084 accidents that were recorded by police subject to misclassification errors. The subsample consists of 1796 accidents that were recorded by both imprecise police reports and precise hospital interviews. By the double sampling design, the subsample was randomly selected from the main sample. Thus, the subsample and the main sample have independent and identical distributions.

The main sample and the subsample in Table 1 were categorized by four variables X, X', Y, and Y', where X and Y denote precise personal survey for seat belt usages and driver injuries, and X' and Y' denote im-

Table 1 Data of Highway Safety Research

		Main sample		Subsample			
				$X' = $ Yes		$X' = $ No	
Y'	Y	$X' = $ Yes	$X' = $ No	$X = $ Yes	$X = $ No	$X = $ Yes	$X = $ No
Yes	Yes	1996	13,562	17	3	10	258
	No			3	4	4	25
No	Yes	7,151	58,175	16	3	25	194
	No			100	13	107	1,014

Source: Hochberg (1977).

precise police reports for them. We denote the main sample and the sub-sample by:

$$n = \{n_{+j+l} | j \in \{\text{Yes}, \text{No}\}, l \in \{\text{Yes}, \text{No}\}\},$$

$$m = \{m_{ijkl} | i \in \{\text{Yes}, \text{No}\}, j \in \{\text{Yes}, \text{No}\}, k \in \{\text{Yes}, \text{No}\}, l \in \{\text{Yes}, \text{No}\}\}.$$

For these data, we assume that the main sample data and the sub-sample data have independent and identical multinomial distributions with:

$$\theta_{XX'YY'} = \{p_{ijkl} | i \in \{\text{Yes}, \text{No}\}, j \in \{\text{Yes}, \text{No}\}, k \in \{\text{Yes}, \text{No}\},$$
$$l \in \{\text{Yes}, \text{No}\}\},$$

where $p_{ijkl} = \Pr(X = i, X' = j, Y = k, Y' = l)$. Thus, each of observed data n and m has:

$$f(m | \theta_{XX'YY'}) \propto \prod_{i,j,k,l} p_{ijkl}^{m_{ijkl}},$$

$$f(n | \theta_{X'Y'}) \propto \prod_{j,l} p_{+j+l}^{n_{+j+l}},$$

where $\theta_{X'Y'} = \{p_{+j+l} | j \in \{\text{Yes}, \text{No}\}, l \in \{\text{Yes}, \text{No}\}\}$ and $p_{+j+l} = \Pr(X' = j, Y' = l)$. For this model, the prior distribution of $\theta_{XX'YY'}$ has the Dirichlet distribution with hyperparameters $\alpha_{XX'YY'} = \{\alpha_{ijkl} | i \in \{\text{Yes}, \text{No}\}, j \in \{\text{Yes}, \text{No}\}, k \in \{\text{Yes}, \text{No}\}, l \in \{\text{Yes}, \text{No}\}\}$:

$$\pi(\theta_{XX'YY'} | \alpha_{XX'YY'}) \propto \prod_{i,j,k,l} p_{ijkl}^{\alpha_{ijkl}-1}.$$

Utilizing the subsample m in Table 1 as hyperparameters $\alpha_{XX'YY'}$, the DA algorithm that augments the incomplete data n is given by the following scheme:

Initialization: Setting the initial distribution

$$\pi^{(0)}(\theta_{XX'YY'} \mid n, m) = \pi(\theta_{XX'YY'} \mid \alpha_{XX'YY'}).$$

Imputation step: To obtain the imputed data of n such that:

$$\tilde{n} = \left\{ \tilde{n}_{ijkl} | i \in \{\text{Yes, No}\}, j \in \{\text{Yes, No}\}, k \in \{\text{Yes, No}\}, \right.$$

$$\left. l \in \{\text{Yes, No}\}, n_{+j+l} = \sum_{i,k} \tilde{n}_{ijkl}, \tilde{n}_{ijkl} \geq 0 \right\},$$

repeating the following steps for $l = 1, \ldots, L$.

1. Generate $\theta_{XX'YY'}^{*}$ from $\pi^{(t-1)}(\theta_{XX'YY'}|n,m)$.
2. Generate $\tilde{n}^{(l)}$ from the predictive distribution, which has the conditional multinomial distribution, given n, with the density:

$$f(\tilde{n}|n, \{p_{i,k|j,l}^{*}\}) \propto \prod_{i,j,k,l} p_{i,k|j,l}^{*\tilde{n}_{ijkl}},$$

where $p_{i,k|j,l}^{*} = p_{ijkl}^{*}/p_{+j+l}^{*}$.

Posterior step: Update the current approximation $\pi^{(t-1)}(\theta_{XX'YY'}| n, m)$, given $\tilde{n}^{(l)}$ for n, by the Monte Carlo method:

$$\pi^{(t)}(\theta_{XX'YY'}|n, m) = \frac{1}{L} \sum_{l=1}^{L} \pi(\theta_{XX'YY'}|\tilde{n}^{(l)}, m),$$

where:

$$\pi(\theta_{XX'YY'}|\tilde{n}^{(l)}, m) \propto \prod_{i,j,k,l} p_{ijkl}^{m_{ijkl}+\tilde{n}_{ijkl}^{(l)}-1}.$$

After convergence of the DA algorithm, we estimate the posterior distribution of the marginal probabilities of X and Y:

$$\theta_{XY} = \{p_{i+k+}|i \in \{\text{Yes, No}\}, k \in \{\text{Yes, No}\}\},$$

where $p_{i+k+} = \Pr(X = i, Y = k)$. To examine the performance of the DA algorithm, Kuroda and Geng (2002) compared with the exact Bayes

Table 2 Estimates and Their SDs of θ_{XY}

		Exact Bayes	DA	Fisher scoring	EM
		Posterior	Posterior		
X	Y	mean±SD	mean±SD	Estimate±SD	
Yes	Yes	0.0397±0.0043	0.0389±0.0041	0.0394±0.0045	0.0394
	No	0.1293±0.0065	0.1311±0.0073	0.1190±0.0076	0.1294
No	Yes	0.2558±0.0079	0.2577±0.0078	0.2563±0.0103	0.2559
	No	0.5752±0.0093	0.5722±0.0092	0.5870±0.0116	0.5752

estimate the EM algorithm and the Fisher scoring algorithm. Table 2 shows the estimates and the standard deviations (SDs) of θ_{XY} obtained by the DA algorithm, and the exact Bayesian calculation, the Fisher scoring algorithm, and the EM algorithm. The estimates using the DA algorithm can be found from simulated samples 1,000,000 after a burn-in sample of 10,000 in two chains. The exact values of estimates of θ_{XY} using the Bayesian calculation are given by Geng and Asano (1989) who assumed the Jeffreys noninformative prior. The estimation using the Fisher scoring algorithm was carried out with l_{EM} developed by Vermunt (1997). From these numerical results, it can be seen that the DA algorithm has the equivalent performance of the EM and the Fisher scoring algorithm in comparison with these estimates and SDs.

4 THE GIBBS SAMPLING ALGORITHM

The Gibbs Sampling algorithm is a multivariate extension of the DA algorithm. The GS algorithm is available for statistical models where it is difficult to draw samples from a joint distribution but easy to generate samples from the set of full conditional distributions of the joint distribution. Both algorithms are closely related and are usually applied to the Bayesian missing data analysis.

Consider the GS algorithm for a model with two parameters $\theta = (\theta_1, \theta_2)$. Let $\pi(\theta)$ denote the joint distribution of θ, and let $\pi(\theta_1|\theta_2)$ and $\pi(\theta_2|\theta_1)$ denote the conditional distributions. Then the marginal distribution $\pi(\theta_1)$ can be calculated from

$$\pi(\theta_1) = \int \pi(\theta_1, \theta_2) d\theta_2 = \int \pi(\theta_1|\theta_2)\pi(\theta_2) d\theta_2, \tag{10}$$

and the marginal distribution $\pi(\theta_2)$ can be also derived from:

$$\pi(\theta_2) = \int \pi(\theta_1, \theta_2)d\theta_1 = \int \pi(\theta_2|\theta_1)\pi(\theta_1)d\theta_1. \qquad (11)$$

Similar to the DA algorithm, by substituting Eq. (11) into Eq. (10) and by interchanging the integration order, we have:

$$\pi(\theta_1) = \int \pi(\theta_1|\theta_2)\left(\int \pi(\theta_2|\phi_1)\pi(\phi_1)d\phi_1\right)d\theta_2$$

$$= \int K(\phi_1, \theta_1)\pi(\phi_1)d\phi_1,$$

where

$$K(\phi_1, \theta_1) = \int \pi(\theta_1|\theta_2)\pi(\theta_2|\phi_1)d\theta_2 \qquad (12)$$

and $K(\phi_1, \theta_1)$ is the transition kernel. Specifying the initial values $(\theta_1^{(0)}, \theta_2^{(0)})$, the GS algorithm produces the sequence of samples:

$$\theta_1^{(0)}, \theta_2^{(0)}, \theta_1^{(1)}, \theta_2^{(1)}, \ldots,$$

from $\pi(\theta_1|\theta_2^{(t-1)})$ and $\pi(\theta_2|\theta_1^{(t)})$ for $t = 1, 2, \ldots$, which forms Eq. (12). For $\pi(\theta_2)$, the same argument is applicable. After a sufficient large number of the sample generation, the distribution of $\theta^{(t)}$ converges to a stationary distribution $\pi(\theta)$. Then the GS algorithm can simulate the samples from $\pi(\theta)$.

For a general multiparameter model, we formulate the GS algorithm. Suppose that a parameter vector θ is partitioned into D components:

$$\theta = (\theta_1, \theta_2, \ldots, \theta_D),$$

where each θ_d is a scalar or a vector. Let $\pi(\theta)$ denote the joint distribution of θ, and let $\pi(\theta_d|\theta_{-d})$ denote the conditional distribution of θ_d given θ_{-d}, where θ_{-d} refers to $\{\theta_c|c \neq d, c \in \{1, \ldots, D\}\}$. Denoting $\phi = (\phi_1, \ldots, \phi_D)$, the transition kernel $K(\phi, \theta)$ is constructed by the set of full conditional distributions and can be expressed by:

$$K(\phi, \theta) = \prod_{d=1}^{D} \pi(\theta_d|\theta_1, \ldots, \theta_{d-1}, \phi_{d+1}, \ldots, \phi_D).$$

The form of $K(\phi,\theta)$ indicates that the limiting distribution of $\theta^{(t)}$ is $\pi(\theta)$. Thus, in order to find $\pi(\theta)$ the GS algorithm iterates the following successive generation:

Step 1: Initialize the parameter vector $\theta^{(0)} = (\theta_1^{(0)}, \ldots, \theta_D^{(0)})$.
Step 2: Generate samples from each of conditional distributions:

$$\theta_1^{(t)} \sim \pi(\theta_1 | \theta_2^{(t-1)}, \ldots, \theta_D^{(t-1)}),$$

$$\theta_2^{(t)} \sim \pi(\theta_2 | \theta_1^{(t)}, \theta_3^{(t-1)}, \ldots, \theta_D^{(t-1)}),$$

$$\vdots$$

$$\theta_D^{(t)} \sim \pi(\theta_D | \theta_1^{(t)}, \ldots, \theta_{D-1}^{(t)}).$$

Step 3: Repeat Step 2 until convergence is reached.

For the discrete distribution, Geman and Geman (1984) proved that the following results hold:

Result 1 (*Convergence*): $\theta^{(t)} = (\theta_1^{(t)}, \ldots, \theta_D^{(t)}) \sim \pi(\theta)$ as $t \to \infty$.
Result 2 (*Rate*): The distribution of $\theta^{(t)}$ converges to $\pi(\theta)$ at an exponential rate in t using L_1 norm.

Next consider the inference of marginal distribution $\pi(\theta_d)$ associated with the density estimation and the calculation of the mean and the variance. After achieving convergence, we produce the samples of θ such that:

$$\theta^{(1)} = (\theta_1^{(1)}, \theta_2^{(1)}, \ldots, \theta_D^{(1)}),$$

$$\theta^{(2)} = (\theta_1^{(2)}, \theta_2^{(2)}, \ldots, \theta_D^{(2)}),$$

$$\vdots$$

$$\theta^{(T)} = (\theta_1^{(T)}, \theta_2^{(T)}, \ldots, \theta_D^{(T)}).$$

Then the density of $\pi(\theta_d)$ can be calculated by using a *kernel density estimator* based on T samples of θ_d. With respect to the mean and the variance of θ_d, these estimates can be directly obtained by the Monte Carlo method:

$$\bar{\theta}_d = \frac{1}{T} \sum_{t=1}^{T} \theta_d^{(t)}, \quad \mathrm{Var}[\theta_d] = \frac{1}{T} \sum_{t=1}^{T} (\theta_d^{(t)} - \bar{\theta}_d)^2. \tag{13}$$

Increasing the value of T, these Monte Carlo estimates are more accurate. When the calculation of the mean and the variance of θ_d is available in the closed forms for $\pi(\theta_d|\theta_{-d})$, the precision of these estimates can be greatly improved by introducing the idea of *Rao–Blackwellization* by Gelfand and Smith (1990) and Liu et al. (1994). The Rao–Blackwellized estimate for the mean of θ_d given $\theta_{-d}^{(t)}$, for $t = 1, \ldots, T$, takes the form:

$$\tilde{\theta}_d = \frac{1}{T} \sum_{t=1}^{T} E[\theta_d|\theta_{-d}^{(t)}] \tag{14}$$

and then is unbiased for $E[\theta_d]$, because:

$$\int E[\theta_d|\theta_{-d}]\pi(\theta_{-d})\mathrm{d}\theta_{-d} = E[\theta_d].$$

The Rao–Blackwell theorem states that it holds $\mathrm{Var}[\bar{\theta}_d] \geq \mathrm{Var}[\tilde{\theta}_d]$, so that $\tilde{\theta}_d$ is more efficient than $\bar{\theta}_d$ in terms of mean square error (MSE). Thus, we can see that the Rao–Blackwellization yields more precious estimates of θ_d than the direct estimation, such as Eq. (13), without increasing the number of iterations of the GS algorithm. The technique of the Rao–Blackwellization can also be to applied to the density estimation of $\pi(\theta_d)$. Because:

$$\pi(\theta_d) = \int \pi(\theta_d|\theta_{-d})\pi(\theta_{-d})\mathrm{d}\theta_{-d},$$

the Rao–Blackwellized density estimation can be performed by calculating:

$$\pi(\theta_d) = \frac{1}{T} \sum_{t=1}^{T} \pi(\theta_d|\theta_{-d}^{(t)}). \tag{15}$$

Naturally, it requires that the density of $\pi(\theta_d)$ is given by the closed form for $\pi(\theta_d|\theta_{-d})$ We also notice that the *Posterior step* of the DA algorithm updates the posterior distributions of interesting parameters by using the Rao–Blackwellization of Eq. (15).

Example: Genetic Linkage Model (Continued)

Gelfand and Smith (1990) extended to the genetic linkage example and applied the GS algorithm to inference of interesting parameters. Assume that the observed data:

$$x = (x_1, x_2, x_3, x_4, x_5) = (14, 1, 1, 1, 5)$$

have the multinomial distribution with parameters:

$$\theta = \left(\frac{1}{4}p_1 + \frac{1}{8}, \frac{1}{4}p_1, \frac{1}{4}p_2, \frac{1}{4}p_2 + \frac{3}{8}, \frac{1}{4}(1 - p_1 - p_2)\right).$$

For this multinomial model, we split x_1 and x_4 into two cells such that $x_1 = y_1 + y_2$ and $x_4 = y_5 + y_6$, and set $x_2 = y_3$, $x_3 = y_4$, and $x_5 = y_7$. Denoting the augmented data of x as y, the likelihood function of y has the multinomial distribution with parameters:

$$\theta' = \left(\frac{1}{4}p_1, \frac{1}{8}, \frac{1}{4}p_1, \frac{1}{4}p_2, \frac{1}{4}p_2, \frac{3}{8}, \frac{1}{4}(1 - p_1 - p_2)\right)$$

and has the density:

$$f(y|\theta') \propto p_1^{y_1+y_3} p_2^{y_4+y_5} (1 - p_1 - p_2)^{y_7}. \tag{16}$$

For (p_1, p_2), we assume that the prior distribution has the Dirichlet distribution with hyperparameters $\alpha = (\alpha_1, \alpha_2, \alpha_3)$ and has the density:

$$\pi(p_1, p_2|\alpha) \propto p_1^{\alpha_1-1} p_2^{\alpha_2-1} (1 - p_1 - p_2)^{\alpha_3-1}. \tag{17}$$

Multiplying the Dirichlet prior density (Eq. (17)) by the multinomial likelihood density (Eq. (16)) produces the posterior density:

$$\pi(p_1, p_2|y) \propto \sum_{y_1=0}^{x_1} \frac{x_1!}{y_1!(x_1 - y_1)!} \sum_{y_5=0}^{x_4} \frac{x_4!}{y_5!(x_4 - y_5)!} p_1^{\alpha_1+y_1+y_3-1}$$
$$\times p_2^{\alpha_2+y_4+y_5-1} (1 - p_1 - p_2)^{\alpha_3+y_7-1}.$$

Viewing (y_1, y_5) as missing data, we require to simulate the samples of (p_1, p_2, y_1, y_5) using the GS algorithm. Then each of the full conditional distributions of p_1, p_2, y_1, and y_5 is specified as follows:

$$\frac{p_1}{1 - p_2} \sim \text{beta}(\alpha_1 + y_1 + y_3, \alpha_3 + y_7), \tag{18}$$

$$\frac{p_2}{1 - p_1} \sim \text{beta}(\alpha_2 + y_4 + y_5, \alpha_3 + y_7), \tag{19}$$

$$y_1 \sim \text{binomial}\left(x_1, \frac{2p_1}{2p_1 + 1}\right), \tag{20}$$

$$y_5 \sim \text{binomial}\left(x_4, \frac{2p_2}{2p_2 + 3}\right), \tag{21}$$

The GS algorithm is performed by generating each of the samples from the full conditional distributions (Eqs. (18) Eqs. (19) Eqs. (20) Eqs. (21)) iteratively.

In this example, we set $\alpha = (1,1,1)$. After 200 burn-in samples, we simulated 10,000 samples and apply the *direct* and the *Rao–Blackwellized* estimation to obtain estimates of (p_1,p_2). Table 3 shows the estimates of the posterior means and variances, and their standard errors (SEs) in parentheses. The exact values of (p_1,p_2) were also calculated by Gelfand and Smith (1990) using "exact numerical methods" in Naylor and Smith (1982). The estimates by the direct estimation are calculated from Eq. (13). The Rao–Blackwellizied estimation gives the posterior means of (p_1,p_2) by the forms:

$$E[p_1|y] = (1 - p_2)\frac{\alpha_1 + y_1 + y_3}{\alpha_1 + \alpha_3 + y_1 + y_3 + y_7},$$

$$E[p_2|y] = (1 - p_1)\frac{\alpha_2 + y_4 + y_5}{\alpha_2 + \alpha_3 + y_4 + y_5 + y_7}.$$

In order to obtain these posterior variances, it requires a less complicated calculation and has the form:

$$\text{Var}[p_i|y] = \text{Var}[E[p_i|y]] + E[\text{Var}[p_i|y]],$$

for $i = 1,2$. From these numerical results, all of the SEs of the Rao–Blackwellized estimates are about half of the direct estimates. The histograms and the trace plots of (p_1,p_2) in Figs. 1 and 2 also indicate that the Rao–Balckwellized estimation can give more efficient estimates from the viewpoint of the reduction of the SEs. These results lead to the conclusion that the Rao–Blackwellized estimation provides the improvement of estimates.

Table 3   Estimates of p_1 and p_2 in the Genetic Linkage Model

	Exact values	GS	
		Direct estimate	Rao–Blackwell estimate
$E[p_1|y]$	0.5199	0.5204 (0.0013)	0.5120 (0.0008)
$E[p_2|y]$	0.1232	0.1234 (0.0008)	0.1230 (0.0004)
$\text{Var}[p_1|y]$	0.0178	0.0178 (0.0006)	0.1777 (0.0003)
$\text{Var}[p_2|y]$	0.0066	0.0066 (0.0002)	0.0063 (0.0001)

Source: Gelfand and Smith (1990).

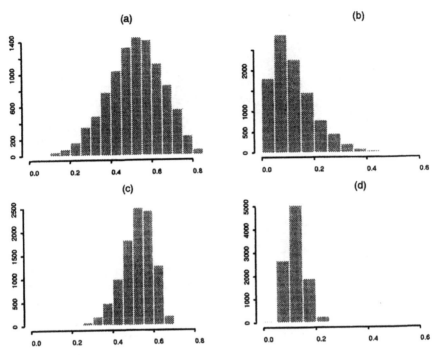

Figure 1 Histogram of 10,000 simulated values of posterior means of p_1 and p_2: (a) direct estimation of p_1; (b) direct estimation of p_2; (c) Rao–Blackwellized estimation of p_1; and (d) Rao–Blackwellized estimation of p_2.

Example: Case–Control Studies with Misclassification Error

Kuroda and Geng (2002) compared the estimates obtained by the DA algorithm with the posterior means using the exact Bayesian calculation given by Geng and Asano (1987) regarding the data from Diamond and Lilienfeld (1962) that reported a case–control study concerning the circumcision status of male partners of woman with and without cervical cancer. In this example, we apply the GS algorithm for this misclassified case–control data. The study sample was categorized by cervical cancer status, X (Case and Control), and self-reported circumcision status, Y' (Yes or No), in the left side of Table 4. In order to gain the information on the degree of misclassification of circumcision status, the supplemental sample concerning the relationship between actual circumcision status, Y (Yes or No), and Y' was gathered from the separate population

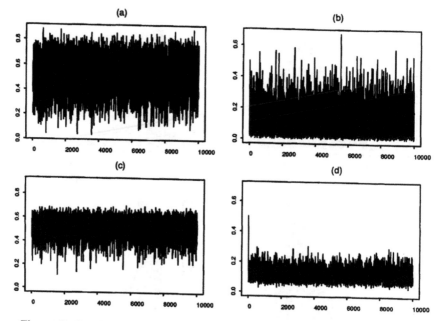

Figure 2 Trace of posterior means of p_1 and p_2: (a) direct estimation of p_1; (b) direct estimation of p_2; (c) Rao–Blackwellized estimation of p_1; and (d) Rao–Blackwellized estimation of p_2.

Table 4 Data from Diamond and Lilienfeld (1962) and Hypothetical Prior Information $\alpha_{XYY'}$

| Y' | Y | Study sample X | | Supplement sample X | Hypothetical prior X | |
		Case	Control	Unknown	Case	Control
Yes	Yes	5	14	37	80	10
	No			19	20	40
No	Yes	95	86	47	40	20
	No			89	10	80

shown in the center of Table 4. We denote the observed data of X and Y', and Y and Y' as:

$$n = \{n_{i+k} | i \in \{\text{Case}, \text{Control}\}, k \in \{\text{Yes}, \text{No}\}\},$$
$$m = \{m_{+jk} | j \in \{\text{Yes}, \text{No}\}, k \in \{\text{Yes}, \text{No}\}\},$$

respectively. Let $p_{ijk} = \Pr(X=i, Y=j, Y'=k)$ and:

$$\theta_{XYY'} = \{p_{ijk} | i \in \{\text{Case}, \text{Control}\}, j \in \{\text{Yes}, \text{No}\}, k \in \{\text{Yes}, \text{No}\}\}.$$

We assume that n and m have the multinomial distributions with parameters $\theta_{XY'}$ and $\theta_{YY'}$, where:

$$\theta_{XY'} = \{p_{i+k} | i \in \{\text{Case}, \text{Control}\}, k \in \{\text{Yes}, \text{No}\}\},$$
$$\theta_{YY'} = \{p_{+jk} | j \in \{\text{Yes}, \text{No}\}, k \in \{\text{Yes}, \text{No}\}\},$$

and $p_{i+k} = \Pr(X=i, Y'=k)$ and $p_{+jk} = \Pr(Y=j, Y'=k)$. For this misclassified multinomial model, we assume that the prior distribution of $\theta_{XYY'}$ has the Dirichlet distribution with hyperparameters $\alpha_{XYY'}$ where:

$$\alpha_{XYY'} = \{\alpha_{ijk} | i \in \{\text{Case}, \text{Control}\}, j \in \{\text{Yes}, \text{No}\}, k \in \{\text{Yes}, \text{No}\}\},$$

and write the prior density:

$$\pi(\theta_{XYY'} | \alpha_{XYY'}) \propto \prod_{i,j,k} p_{ijk}^{\alpha_{ijk}-1}$$

Given n and m, the posterior distribution of θ has the mixture Dirichlet distribution with density:

$$\pi(\theta_{XYY'} | n, m) \propto \prod_{i,k} p_{i+k}^{n_{i+k}} \prod_{j,k} p_{+jk}^{m_{+jk}} \prod_{i,j,k} p_{ijk}^{\alpha_{ijk}-1}$$

$$= \prod_{k} \left\{ \prod_{i} \sum_{\Omega(n)} \frac{n_{i+k}!}{\prod_{j} \tilde{n}_{ijk}!} \prod_{j} \sum_{\Omega(m)} \frac{m_{+jk}!}{\prod_{i} \tilde{m}_{ijk}!} p_{ijk}^{\alpha_{ijk} + \tilde{n}_{ijk} + \tilde{m}_{ijk} - 1} \right\}$$

$$(22)$$

where $\sum_{\Omega(n)}$ denotes the sum over all possible $\{\tilde{n}_{ijk}\}$ under the conditions $\tilde{n}_{ijk} \geq 0$ for all i, j, and k, and $n_{i+k} = \sum_{j} \tilde{n}_{ijk}$, and $\sum_{\Omega(m)}$ denotes the sum over all possible $\tilde{m}_{ijk}$ under the conditions $\tilde{m}_{ijk} \geq 0$ for all i, j, and k, and $m_{+jk} = \sum_{i} \tilde{m}_{ijk}$. The GS algorithm is applied to find $\pi(\theta_{XYY'} | n, m)$ and estimate posterior means and variances.

Geng and Asano (1989) gave the hypothetical prior information shown in the right side of Table 4. We use their prior information as hyperparameters. Let the imputed data of n and m be denoted as:

$$\tilde{n} = \left\{ \tilde{n}_{ijk} | i \in \{\text{Case}, \text{Control}\}, j \in \{\text{Yes}, \text{No}\}, l \in \{\text{Yes}, \text{No}\}, n_{i+k} \right.$$

$$\left. = \sum_j \tilde{n}_{ijk}, \tilde{n}_{ijk} \geq 0 \right\},$$

$$\tilde{m} = \left\{ \tilde{m}_{ijk} | i \in \{\text{Case}, \text{Control}\}, j \in \{\text{Yes}, \text{No}\}, l \in \{\text{Yes}, \text{No}\}, m_{+jk} \right.$$

$$\left. = \sum_i \tilde{m}_{ijk}, \tilde{m}_{ijk} \geq 0 \right\}.$$

Then the predictive distributions of $\tilde{n}$ and $\tilde{m}$ have the multinomial distributions conditionally as n and m, and these densities are given by:

$$f(\tilde{n}|n, p_{j|i,k}) \propto \prod_{i,j,k} p_{j|i,k}^{\tilde{n}_{ijk}},$$

$$f(\tilde{m}|m, \{p_{i|j,k}\}) \propto \prod_{i,j,k} p_{i|j,k}^{\tilde{m}_{ijk}},$$

where $p_{j|i,k} = p_{ijk}/p_{i+k}$ and $p_{i|j,k} = p_{ijk}/p_{+jk}$. For the misclassified multinomial model, the GS algorithm can be described by the following iterative schemes:

Step 1: Initialize the parameters $\theta^{(0)}_{XYY'}$, $\tilde{n}(0)$, and $\tilde{m}(0)$.

Step 2: Generate each of samples from specified conditional distributions:

$$\theta^{(t)}_{XYY'} \sim \pi(\theta_{XYY'}|\tilde{n}^{(t-1)}, \tilde{m}^{(t-1)}),$$

$$\tilde{n}^{(t)} \sim f(\tilde{n}|n, \{p^{(t)}_{j|i,k}\}),$$

$$\tilde{m}^{(t)} \sim f(\tilde{m}|m, \{p^{(t)}_{i|j,k}\}).$$

Step 3: Repeat Step 2 until convergence is reached.

Step 4: Find the posterior distribution of model parameter:

$$\theta_{XY} = \{p_{ij+} | i \in \{\text{Case}, \text{Control}\}, j \in \{\text{Yes}, \text{No}\}\}.$$

In this numerical experiment, we evaluate the precision of the estimates using the GS algorithm in comparison with the exact posterior

Table 5 Posterior Means and SDs and 95% CIs Using the DA Algorithm and the Exact Posterior Means

		Exact Bayes	DA	
X	Y	Posterior means	Posterior mean$\pm$SD	CI (lower–upper)
Case	Yes	0.3794	0.3786$\pm$0.0127	0.3512–0.4017
	No	0.1116	0.1134$\pm$0.0159	0.0838–0.1460
Control	Yes	0.0921	0.0927$\pm$0.0107	0.0737–0.1142
	No	0.4169	0.4152$\pm$0.0113	0.3916–0.4364

Source: Geng and Asano (1989).

means given by Geng and Asano (1989). Table 5 shows the exact values, the posterior means, the SDs, and the posterior 95% credible intervals (CIs) of θ_{XY} obtained from simulated 10,000 samples after a burn-in of 1000 samples. It can be seen that the estimates have approximately three-digit precision for the exact values.

5 THE METROPOLIS–HASTINGS ALGORITHM

The Metropolis–Hastings algorithm is the general MCMC method and is different from the GS algorithm in the procedure to generating samples. For the statistical model, it is hard to generate samples from the set of full conditional distributions of a joint distribution $\pi(\theta)$; the GS algorithm does not work well. Then the M–H algorithm is suitable because of drawing of samples not from $\pi(\theta)$ but from a candidate density that dominates or blankets $\pi(\theta)$. Furthermore, the M–H algorithm is applicable to the situation where $\pi(\theta)$ does not have a standard statistical distribution.

Suppose we have a transition density $q(\theta|\phi)$ to move from $\theta^{(t)} = \phi$ to $\theta^{(t+1)} = \theta$, where $\int q(\theta|\phi)d\theta = 1$. As described in Sec. 2, if $q(\theta|\phi)$ satisfies time reversibility such that:

$$\pi(\phi)q(\theta|\phi) = \pi(\theta)q(\phi|\theta) \tag{23}$$

for all ϕ and θ, then $\pi(\theta)$ is the stationary distribution of the Markov chain $\{\theta^{(t)}|t \geq 0\}$. But it will not be most likely to satisfy the reversible condition. Thus, our object is how the M–H algorithm finds $q(\theta|\phi)$ with the reversibility. Chib and Greenberg (1995) described in detail the process of deriving the transition kernel of the M–H algorithm. We will provide the formulation of the kernel according to their exposition.

Assume that the following inequality holds for any ϕ and θ:

$$\pi(\phi)q(\theta|\phi) > \pi(\theta)q(\phi|\theta). \tag{24}$$

Then Eq. (24) means that θ can be often generated from $q(\theta|\phi)$, but ϕ can be generated from $q(\phi|\theta)$, rarely. To correct this inequality, we introduce a probability $0 \le \alpha(\phi,\theta) \le 1$, which refers to the *acceptance probability* moving from $\theta^{(t)} = \phi$ to $\theta^{(t+1)} = \theta$. For $\phi \ne \theta$, the transition density $K(\phi,\theta)$ is given by:

$$K(\phi|\theta) = q(\theta|\phi)\alpha(\phi,\theta). \tag{25}$$

Then Eq. (24) shows that it is not enough to generate ϕ from $q(\phi|\theta)$, so that $\alpha(\theta,\phi)$ should be set to one that is the upper limit. Thus:

$$\begin{aligned}
\pi(\phi)q(\theta|\phi)\alpha(\phi,\theta) &= \pi(\theta)q(\phi|\theta)\alpha(\theta,\phi) \\
&= \pi(\theta)q(\phi|\theta),
\end{aligned} \tag{26}$$

and we have:

$$\alpha(\phi,\theta) = \frac{\pi(\theta)q(\phi|\theta)}{\pi(\phi)q(\theta|\phi)}. \tag{27}$$

If the inequality in Eq. (24) is reverse, we can set $\alpha(\phi,\theta) = 1$ and obtain $\alpha(\theta,\phi)$ by the same way. Introducing $\alpha(\phi,\theta)$ and $\alpha(\theta,\phi)$ for $\theta \ne \phi$, we can find $K(\phi|\theta)$ that satisfies the reversible condition. Then $\alpha(\phi,\theta)$ must be set:

$$\alpha(\phi,\theta) = \begin{cases} \min\left[\dfrac{\pi(\theta)q(\phi|\theta)}{\pi(\phi)q(\theta|\phi)}, 1\right], & \text{if } \pi(\phi)q(\theta|\phi) > 0, \\ 1, & \text{if } \pi(\phi)q(\theta|\phi) = 0. \end{cases} \tag{28}$$

Next we derive the probability $r(\phi)$ moving from $\theta^{(t)} = \phi$ to $\theta^{(t+1)} = \phi$. For this transition, there are two events: one is to move back to the same point and another is not to move. Then $r(\phi)$ is given by:

$$r(\phi) = 1 - \int K(\phi,\theta)d\theta = 1 - \int q(\theta|\phi)\alpha(\phi,\theta)d\theta. \tag{29}$$

Thus, the transition kernel $K(\phi,\cdot)$ of the M–H algorithm is given by:

$$\begin{aligned}
K(\phi,A) &= \int_A q(\theta|\phi)\alpha(\phi,\theta)d\theta + I_A(\phi)r(\phi) \\
&= \int_A q(\theta|\phi)\alpha(\phi,\theta)d\theta + I_A(\phi)\left[1 - \int q(\theta|\phi)\alpha(\phi,\theta)d\theta\right],
\end{aligned} \tag{30}$$

where A is any subset of parameter space, and $I_A(\phi) = 1$ if $\phi \in A$ and zero otherwise. Then:

$$\int K(\phi, A)\pi(\phi)d\phi = \int \left(\int_A q(\theta|\phi)\alpha(\phi, \theta)d\theta + I_A(\phi)r(\phi) \right)\pi(\phi)d\phi$$

$$= \int_A \left(\int q(\theta|\phi)\alpha(\phi, \theta)\pi(\phi)d\phi \right)d\theta + \int_A r(\phi)\pi(\phi)d\phi$$

$$= \int_A \left(\int q(\phi|\theta)\alpha(\theta, \phi)\pi(\theta)d\phi \right)d\theta + \int_A r(\phi)\pi(\phi)d\phi$$

$$= \int_A \left(\int q(\phi|\theta)\alpha(\theta, \phi)d\phi \right)\pi(\theta)d\theta + \int_A r(\phi)\pi(\phi)d\phi$$

$$= \int_A (1 - r(\theta))\pi(\theta)d\theta + \int_A r(\phi)\pi(\phi)d\phi$$

$$= \int_A \pi(\theta)d\theta,$$

so that there exists the stationary distribution $\pi(\theta)$. We note that the GS algorithm can be regarded as the M–H algorithm for the case that $q(\theta|\phi) = \pi(\theta|\phi)$ and $\alpha(\phi,\theta) = 1$ for all ϕ and θ. Thus, we see that the GS algorithm converges to the true joint distribution $\pi(\theta)$.

In order to generate samples from $\pi(\theta)$, the M–H algorithm can be accomplished by the following procedures:

Step 1: Set the initialize the parameter $\theta^{(0)}$.
Step 2: Repeat the following steps for $t = 1, 2, \ldots, T$.

    1. Generate a candidate sample θ from $q(\theta|\phi)$, where $\theta^{(t)} = \phi$.

    2. Calculate the acceptance probability $\alpha(\phi,\theta)$ from

$$\alpha(\phi, \theta) = \min\left[\frac{\pi(\theta)q(\phi|\theta)}{\pi(\phi)q(\theta|\phi)}, 1 \right].$$

    3. Take

$$\theta^{(t+1)} = \begin{cases} \theta, & \text{with probability } \alpha(\phi, \theta), \\ \phi, & \text{with probability } 1 - \alpha(\phi, \theta). \end{cases}$$

Step 3: Obtain the samples $\{\theta^{(1)}, \ldots, \theta^{(T)}\}$.

The choice of $q(\theta|\phi)$ is critical in applying the M–H algorithm to statistical models because it is closely related to the efficiency of the algorithm and the convergence speed. Thus, it is required to select a suitable candidate density $q(\theta|\phi)$. We describe four types of the candidate density.

A first type is the *symmetrical* type, where the candidate density is symmetrical such that $q(\theta|\phi) = q(\phi|\theta)$, as proposed by Metropolis et al. (1953). Then we have:

$$\alpha(\phi, \theta) = \min\left[\frac{\pi(\theta)}{\pi(\phi)}, 1\right].$$

A second type is the *random walk* type such that $q(\theta|\phi) = q(\theta-\phi)$. Then the candidate value θ is obtained from $\theta = \phi + \varepsilon$, where ε is generated from $q(\varepsilon)$. Note that the random walk type is identical to the symmetrical type when $q(\varepsilon) = q(-\varepsilon)$.

A third type is the *independent* type. Hastings (1970) suggested this type M–H algorithm. Then the generation of θ is independent of ϕ, so that:

$$\alpha(\phi, \theta) = \min\left[\frac{\pi(\theta)q(\phi)}{\pi(\phi)q(\theta)}, 1\right].$$

A fourth type is the *Accept–Reject* (A–R) type, which is an extension of the A–R algorithm and was developed by Tierney (1994). Suppose that a constant $c > 0$ is known, and $h(\theta)$ is a probability density function and is easy to produce samples. Then the A–R-type M–H algorithm does not assume that it is always necessary to hold $\pi(\theta) < ch(\theta)$. Define the set $C = \{\theta|\pi(\theta) < ch(\theta)\}$ and denote the complement set of C as C^c. Based on the idea of the A–R algorithm, we can obtain:

$$q(\theta|\phi) \propto \pi(\theta)/c, \quad \text{if } \theta \in C,$$

$$\propto h(\theta), \quad \text{if } \theta \in C^c.$$

Then, because θ is drawn independently of ϕ, it holds $q(\theta|\phi) = q(\theta)$. For the A–R-type M–H algorithm, we find $\alpha(\phi,\theta)$ that $q(\theta|\phi)\alpha(\phi,\theta)$ satisfies reversibility. Because ϕ and θ are in C or C^c, there are four possible cases: (a) $\phi \in C$, $\theta \in C$; (b) $\phi \in C^c$, $\theta \in C$; (c) $\phi \in C$, $\theta \in C^c$; and (d) $\phi \in C^c$, $\theta \in C^c$. For these cases:

$$\alpha(\phi, \theta) = \min\left[\frac{\pi(\theta)q(\phi|\theta)}{\pi(\phi)q(\theta|\phi)}, 1\right]$$

$$
= \begin{cases}
1, & \text{if } \phi \in C, \theta \in C \text{ or } \phi \in C, \theta \in C^c, \\[2mm]
\dfrac{ch(\phi)}{\pi(\phi)}, & \text{if } \phi \in C^c, \theta \in C, \\[2mm]
\min\left[\dfrac{\pi(\theta)h(\phi)}{\pi(\phi)h(\theta)}\right], & \text{if } \phi \in C^c, \theta \in C^c.
\end{cases}
$$

With the detailed derivation of $\alpha(\phi,\theta)$ for the A–R-type, refer to Chib and Greenberg (1995).

Example: Genetic Linkage Model (Continued)

We apply the M–H algorithm to estimating p for the genetic linkage model illustrated in Sec. 3. For this genetic linkage example, we set:

$$
\pi(\theta) \propto f(x|\theta) \propto (2+p)^{x_1}(1-p)^{x_2+x_3}p^{x_4},
$$

and generate candidate values from the uniform distribution $U(0,1)$, that is, $q(\cdot|p) = 1$ on $[0,1]$. Then the M–H algorithm is described by the following scheme:

Step 1: Initialize the parameter $p^{(0)}$.
Step 2: Repeat the following steps for $t = 1, 2, \ldots, T$.

     1.   Generate a candidate sample ϕ from $q(\phi|p^{(t)})$.
     2.   Calculate $\alpha(p^{(t)},\phi)$ from

$$
\alpha(p^{(t)}, \phi) = \min\left[\frac{\pi(\phi)}{\pi(p^{(t)})}, 1\right].
$$

     3.   Take

$$
p^{(t+1)} = \begin{cases}
\phi, & \text{with probability } \alpha(p^{(t)}, \phi) \\
p^{(t)}, & \text{with probability } 1 - \alpha(p^{(t)}, \phi).
\end{cases}
$$

Step 3: Obtain the samples $\{p^{(1)}, \ldots, p^{(T)}\}$.

After deleting the first 400 observations as burn-in sample, we estimate p using 2000 simulated samples. Then the estimate of p is 0.6237 and the variance is 0.0025. As illustrated in Sec. 3, these estimates are similar to those using the DA algorithm.

Example: Relationship Between the Month of Birthday and Deathday

Table 6 summarizes the month of birthday and deathday for 82 descendants for Queen Victoria by the two-way contingency table. In order to test

Table 6 Relationship Between Birthday and Deathday

Month of birth	Month of death												
	January	February	March	April	May	June	July	August	September	October	November	December	
January	1	0	0	0	1	2	0	0	1	0	1	0	6
February	1	0	0	1	0	0	0	0	0	1	0	2	5
March	1	0	0	0	2	1	0	0	0	0	0	1	5
April	3	0	2	0	0	0	1	0	1	3	1	1	12
May	2	1	1	1	1	1	1	1	1	1	1	0	12
June	2	0	0	0	1	0	0	0	0	0	0	0	3
July	2	0	2	1	0	0	0	0	1	1	1	2	10
August	0	0	0	3	1	0	1	0	0	1	0	2	7
September	0	0	0	1	1	0	0	0	0	0	1	0	3
October	1	1	1	2	0	0	1	0	0	1	1	0	7
November	0	1	1	1	2	0	1	2	0	1	1	0	9
December	0	1	1	0	0	0	1	0	0	0	0	0	3
	13	4	7	10	8	4	5	3	4	9	7	8	82

Source: Andrews and Herzberg (1985).

the hypothesis of association between birthday and deathday, the χ^2 test for independence is usually applied. Then the χ^2 test statistic for independence is 115.6 on 121 df. In general, in the case that the minimal cell is less than five, it is well known that the χ^2 approximation is poor and the exact calculation such as the Fisher exact test is preferable. However, the Fisher exact test is computationally infeasible for the case that there exist an enormous number of enumeration patterns for the contingency table subject to fixed row sums and column sums. In this example, using the random walk type of M–H algorithm, we calculate the cumulative probability of χ^2_{121} (115.6).

Let n_{ij} and p_{ij} denote the cell frequency and probability for i-th category of the month of birth and j-th category of the month of death, respectively. Let n_{i+} and n_{+j} denote the row sum of i-th category and the column sum of j-th category. We write the set of cell frequencies, the row, and the column sums by $n = \{n_{ij} | 1 \le i,\ j \le 12\}$, $n_r = \{n_{i+} | 1 \le i \le 12\}$, and $n_c = \{n_{+j} | 1 \le j \le 12\}$. Then the conditional distribution of n fixed by n_r and n_c has the density:

$$
\pi(n) = \frac{\displaystyle\prod_{i=1}^{12} \frac{n_{i+}!}{\displaystyle\prod_{j=1}^{12} n_{ij}!} \prod_{i,j=2}^{12} \left(\frac{p_{11}p_{ij}}{p_{i1}p_{j1}}\right)^{n_{ij}}}{\displaystyle\sum_{\Omega(m)} \prod_{i=1}^{12} \frac{n_{i+}!}{\displaystyle\prod_{i,j=1}^{12} m_{ij}!} \prod_{i,j=2}^{12} \left(\frac{p_{11}p_{ij}}{p_{i1}p_{j1}}\right)^{m_{ij}}},
$$

where $\Omega(m) = \left\{ m_{ij} \middle| n_{i+} = \sum_j m_{ij}, n_{+j} = \sum_i m_{ij}, m_{ij} \ge 0 \right\}$ and $\sum_{\Omega(m)}$ is the sum over all $m_{ij} \in \Omega(m)$. Under the assumption of the independence model with $(p_{11}p_{ij})/(p_{i1}p_{j1}) = 1$, the M–H algorithm is performed by iterating the following steps for $t = 0, 1, \ldots, T$:

Step 1: Choose a pair $r = (i_1, i_2)$ of rows and a pair $c = (j_1, j_2)$ of columns at random and obtain four cells $n_{rc} = (n_{i_1 j_1}, n_{i_1 j_2}, n_{i_2 j_1}, n_{i_2 j_2})$.

Step 2: Generate the candidate values $m_{rc} = (m_{i_1 j_1}, m_{i_1 j_2}, m_{i_2 j_1}, m_{i_2 j_2})$ from

$$
m_{i_1 j_1} = n_{i_1 j_1} + \varepsilon_{11}, \quad m_{i_1 j_2} = n_{i_1 j_2} + \varepsilon_{12},
$$
$$
m_{i_2 j_1} = n_{i_2 j_1} + \varepsilon_{21}, \quad m_{i_2 j_2} = n_{i_2 j_2} + \varepsilon_{22},
$$

where

$$\begin{pmatrix} \varepsilon_{11} & \varepsilon_{12} \\ \varepsilon_{21} & \varepsilon_{22} \end{pmatrix} = \begin{pmatrix} 1 & -1 \\ -1 & 1 \end{pmatrix} \; or \; \begin{pmatrix} -1 & 1 \\ 1 & -1 \end{pmatrix}$$

with probability 1/2. Obtain the candidate cell frequencies:

$$m = \{m_{rc}, n \backslash n_{rc}\}.$$

Step 3: Calculate $\alpha(n,m)$ from

$$\alpha(n, m) = \min\left[\frac{\pi(m)}{\pi(n)}, 1\right] = \min\left[\frac{n_{i_1 j_1} n_{i_1 j_2} n_{i_2 j_1} n_{i_2 j_2}}{m_{i_1 j_1} m_{i_1 j_2} m_{i_2 j_1} m_{i_2 j_2}}, 1\right].$$

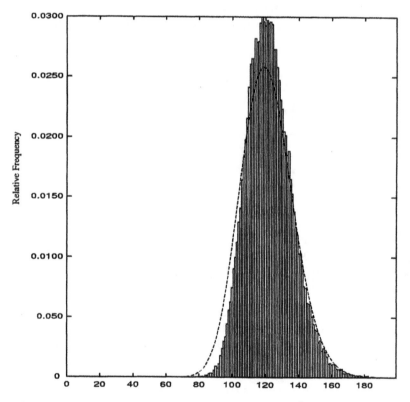

Figure 3   Histogram of 1,000,000 simulated values of χ^2 test statistics.

Step 4: Take

$$m^{(t)} = \begin{cases} m, & \text{with probability } \alpha(n,m), \\ n, & \text{with probability } 1 - \alpha(n,m) \end{cases}$$

Step 5: Calculate the χ^2 test statistic.

Using the random walk type M–H algorithm makes the calculation of $\alpha(n,m)$ easy. From simulated 1,000,000 samples, the estimate of the cumulative probability of $\chi^2_{121} \leq 115.6$ is 0.3181 vs. 0.3785 for χ^2 approximation. Diaconis and Strumfels (1998) calculated the permutation probability of $\chi^2_{121} \leq 115.6$ of 0.3208. The estimate of the cumulative probability using the M–H algorithm is very similar to the permutation probability. The histogram in Fig. 3 shows the relative frequencies for the simulated values of χ^2 statistics obtained by the M–H algorithm. The dashed line also indicates the theoretical χ^2 values with 121 df. The estimate using the M–H algorithm and Fig. 3 shows that the χ^2 approximation is not accurate for this spare contingency table.

BIBLIOGRAPHIC NOTES

There are many excellent books to study deeply the theory and the practical applications of MCMC methods. We will refer to some books. Gilks et al. (1996), Gamerman (1997), and Robert and Casella (1999) cover all areas of theory, implementing methods and practical usages needed in MCMC methods. Schafer (1997) provides the Bayesian inference using the EM and the DA algorithms for incomplete multivariate data whose type are continuous, categorical, and both. All of the computational algorithms are implemented by S language. Congdon (2001) presents the statistical modeling using Bayesian approaches with various real data. In this book, all of the computational programs are coded by WinBUGS, which is the software for the Bayesian analysis using the GS algorithm developed by Spiegelhalter et al. (2000).

We do not deal with the convergence diagnostics for MCMC methods. There exist many methods of convergence diagnostics proposed in the literature. Cowles and Carlin (1996) and Brooks and Roberts (1998) provide the comparative reviews and the numerical illustration of many diagnostic methods. The following four diagnostic methods are easily available in CODA, which is the S function implemented by Best et al. (1995). The convergence diagnostic of Gelman and Rubin (1992) is the variance ratio method based on the analysis of variance that compares with the *within*-variance and the *between*-variance of chains. Their method is appropriate for multiple chains. Geweke (1991) suggests the convergence diagnostic using the standard technique from spectral analysis. His diagnostic is suitable for the case of a single chains. Raftery and Lewis (1992) propose the convergence diagnostic method that finds the minimum number of iterations needed to estimate interesting parameters with desired precision. Their approach is applicable to a single chain and based on two-state Markov chain theory. Heidelberger and Welch (1983) give the procedure based on Brownian bridge theory and using Cramer–von Mises statistics. Their diagnostic detects an initial transient in the sequence of generated single chains.

REFERENCES

1. Andrews, D., Herzberg, A. (1985). *Data*. New York: Springer.
2. Best, N. G., Cowles, M. K., Vines, K. (1995). CODA: convergence diagnosis and output analysis software for Gibbs Sampling output, version 0.30. Technical Report, MRC Biostatistics Unit, University of Cambridge.
3. Brooks, S. P., Roberts, G. O. (1998). Assessing convergence of Markov chain Monte Carlo algorithms. *Stat. Comput.* 8:319–335.
4. Chib, S., Greenberg, E. (1995). Understanding the Metropolis–Hastings algorithm. *Am. Stat.* 49:327–335.
5. Congdon, P. (2001). *Bayesian Statistical Modeling*. Chichester: Wiley.
6. Cowles, M. K., Carlin, B. P. (1996). Markov chain Monte Carlo convergence diagnostics: a comparative review. *J. Am. Stat. Assoc.* 91:883–904.
7. Diaconis, P., Strumfels, B. (1998). Algebraic algorithms for sampling from conditional distributions. *Ann. Stat.* 26:363–397.
8. Diamond, E. L., Lilienfeld, A. M. (1962). Effects of errors in classification and diagnosis in various types of epidemiological studies. *Am. J. Publ. Health.* 52:1137–1144.
9. Gamerman, D. (1997). *Markov Chain Monte Carlo*. London: Chapman & Hall.
10. Gelfand, A. E., Smith, A. F. M. (1990). Sample-based approaches to calculating marginal densities. *J. Am. Stat. Assoc.* 85:398–409.
11. Gelman, A., Rubin, D. B. (1992). Inference from iterative simulation using multiple sequences (with discussion). *Stat. Sci.* 7:457–511.
12. Geman, S., Geman, D. (1984). Stochastic relaxation, Gibbs distributions and the Bayesian restoration of image. *IEEE Trans. Pattern Anal. Mach. Intell.* 6:721–741.
13. Geng, Z., Asano, Ch. (1989). Bayesian estimation methods for categorical data with misclassification. *Commun. Stat.* 8:2935–2954.

14. Geweke, J. (1991). Evaluating the accuracy of sampling-based approaches to the calculation of posterior moments (with discussion). In: Bernerdo, J. M., Berger, J., Dawid, A. P., Smith, A. F. M., eds. *Bayesian Statistics 4*. Oxford: Oxford University Press, pp. 169–193.

15. Gilks, W. R., Richardson, S., Spiegelhalter, D. J. (1996). *Markov Chain Monte Carlo in Practice*. London: Chapman & Hall.

16. Hastings, W. K. (1970). Monte Carlo sampling methods using Markov chains and their applications. *Biometrika* 57:97–109.

17. Heidelberger, P., Welch, P. D. (1983). Simulation run length control in the presence of an initial transient. *Oper. Res.* 31:1109–1144.

18. Hochberg, Y. (1977). On the use of double sampling schemes in analyzing categorical data with misclassification errors. *J. Am. Stat. Assoc.* 72:914–921.

19. Kuroda, M., Geng, Z. (2002). Bayesian inference for categorical data with misclassification errors. *Advanced in Statistics, Combinatorics and Related Areas*. Singapore: World Scientific, pp. 143–151.

20. Liu, J. S., Wong, W. H., Kong, A. (1994). Covariance structure of the Gibbs sampler with applications to the comparisons of estimators and augmentation schemes. *Biometrika* 81:27–40.

21. Louis, T.A. (1982). Finding observed information using the EM algorithm. *J. R. Stat. Soc., B* 44:98–130.

22. Metropolis, N., Rosenbluth, A. W., Rosenbluth, M. N., Teller, A. H., Teller, E. (1953). Equations of state calculations by fact computing machines. *J. Chem. Phys.* 21:1087–1091.

23. Raftery, A. E., Lewis, S. M. (1992). How many iterations in the Gibbs sampler. In: Bernerdo, J. M., Berger, J., Dawid, A. P., Smith, A. F. M., eds. *Bayesian Statistics 4*. Oxford: Oxford University Press, pp. 763–773.

24. Robert, C. R., Casella, G. (1999). *Monte Carlo Statistical Methods*. New York: Springer-Verlag.

25. Schafer, J. L. (1997). *Analysis of Incomplete Multivariate Data*. London: Chapman & Hall.

26. Spiegelhalter, D. J., Thomas, A., Best, N. G. (2000). WinBUGS: Bayesian inference using Gibbs Sampling, version, 1.3. Technical Report, MRC Biostatistics Unit, University of Cambridge.

27. Tanner, M. A., Wong, W. H. (1987). The calculation of posterior distributions by data augmentation. *J. Am. Stat. Assoc.* 82:528–540.

28. Tierney, L. (1994). Markov chains for exploring posterior distributions (with discussion). *Ann. Stat.* 22:1701–1762.

29. Tierney, L. (1995). Introduction to general state-space Markov chain theory. In: Gilks, W. R., Richardson, S., Spiegelhalter, D. J., eds. *Markov Chain Monte Carlo in Practice*. London: Chapman & Hall, pp. 59–74.

30. Vermunt, J. K. (1997). L_{EM}: *A General Program for the Analysis of Categorical Data*. Tilburg University, The Netherlands.

Appendix A: SOLAS™ 3.0 for Missing Data Analysis

1 INTRODUCTION

Missing data are a pervasive problem in data analysis. Missing values lead to less efficient estimates because of the reduced size of the database. In addition, standard complete-data methods of analysis no longer apply. For example, analyses such as multiple regression use only cases that have complete data. Therefore, including a variable with numerous missing values would severely reduce the sample size.

When cases are deleted because one or more variables are missing, the number of remaining cases can be small even if the missing data rate is small for each variable. For example, suppose your data set has five variables measured at the start of the study and monthly for 6 months. You have been told, with great pride, that each variable is 95% complete. If each of these five variables has a random 5% of the values missing, then the proportion of cases that are expected to be complete is: $1-(.95)^{35} = 0.13$. That is, only 13% of the cases would be complete and you would lose 87% of your data.

Missing data also cause difficulties in performing Intent-to-Treat (IT) analyses in randomized experiments. Intent-to-Treat analysis dictates that all cases—complete and incomplete—be included in any analysis. Biases may exist from the analysis of only complete cases if there are systematic differences between completers and dropouts. To select a valid approach for imputing missing data values for any particular variable, it is necessary to consider the underlying mechanism accounting for missing data. Variables in a data set may have values that are missing for different reasons. A laboratory value might be missing because of the following reasons:

> It was below the level of detection.
>
> The assay was not done because the patient did not come in for a scheduled visit.
>
> The assay was not done because the test tube was dropped or lost.
>
> The assay was not done because the patient died, or was lost to follow-up, or other possible causes.

2 OVERVIEWS OF IMPUTATION IN SOLAS™

Imputation is the name given to any method whereby missing values in a data set are filled in with plausible estimates. The goal of any imputation technique is to produce a complete data set that can be analyzed using complete-data inferential methods. The following describes the single and multiple imputation methods available in SOLAS™ 3.0, which are designed to accommodate a range of missing data scenarios in both longitudinal and single-observation study designs.

3 SINGLE IMPUTATION OVERVIEW

Single imputation is a method where each missing value in a data set is filled in with one value to yield one complete data set. This allows standard complete-data methods of analysis to be used on the filled-in data set. SOLAS™ 3.0 provides four distinct methods by which you can perform single imputation: Group Means, Hot-Deck Imputation, Last Value Carried Forward (LVCF), and Predicted Mean Imputation. The single imputation option provides a standard range of traditional imputation techniques useful for sensitivity analysis.

3.1 Group Means

Missing values in a continuous variable will be replaced with the group mean derived from a grouping variable. The grouping variable must be a categorical variable that has no missing data. Of course, if no grouping variable is specified, missing values in the variable to be imputed will be replaced with its overall mean. When the variable to be imputed is categorical, with different frequencies in two or more categories (providing a unique mode), then the modal value will be used to replace missing values in that variable. Note that if there is no unique mode (i.e., if there are equal frequencies in two or more categories) and the variable is nominal, a value will be randomly selected from the categories with the highest frequency. If the variable is ordinal, then the "middle" category will be imputed; or if there are an even number of categories, a value is randomly chosen from the middle two.

3.2 Hot-Deck Imputation

This procedure sorts respondents and nonrespondents into a number of imputation subsets according to a user-specified set of covariates. An imputation subset comprises cases with the same values as those of the user-specified covariates. Missing values are then replaced with values taken from matching respondents (i.e., respondents who are similar with respect to the covariates). If there is more than one matching respondent for any particular nonrespondent, the user has two choices:

The first respondent's value within the imputation subset is used to impute. The reason for this is that the first respondent's value may be closer in time to the case that has the missing value. For example, if cases are entered according to the order in which they occur, there may possibly be some kind of time effect in some studies.

A respondent's value is randomly selected from within the imputation subset. If a matching respondent does not exist in the initial imputation class, the subset will be collapsed by one level starting with the last variable that was selected as a sort variable, or until a match can be found. Note that if no matching respondent is found, even after all of the sort variables have been collapsed, three options are available:

Respecify new sort variables: where the user can specify up to five sort variables.
Perform random overall imputation: where the missing value will be replaced with a value randomly selected from the observed values in that variable.

Do not impute the missing value: where any missing values for which no matching respondent is found will not be imputed.

3.3 Last Value Carried Forward

The Last Value Carried Forward technique can be used when the data are longitudinal (i.e., repeated measures have been taken per subject). The last observed value is used to fill in missing values at a later point in the study and therefore makes the assumption that the response remains constant at the last observed value.

This can be biased if the timing of withdrawal and the and rate of withdrawal are related to the treatment. For example, in the case of degenerative diseases, using the last observed value to impute for missing data at a later point in the study means that a high observation will be carried forward, resulting in an overestimation of the true end-of-study measurement. Longitudinal variables are those variables intended to be measured at several points in time, such as pretest and posttest measurements of an outcome variable made at monthly intervals, laboratory tests made at each visit from baseline, through the treatment period, and through the follow-up period.

For example, if the blood pressures of patients were recorded every month over a period of 6 months, we would refer to this as one longitudinal variable consisting of six repeated measures or periods.

Linear interpolation is another method for filling in missing values in a longitudinal variable. If a missing value has at least one observed value before, and at least one observed value after, the period for which it is missing, then linear interpolation can be used to fill in the missing value. Although this method logically belongs in the LVCF option, for historical reasons, it is only available as an imputation method from within the Propensity Score-Based Method (for further details see the "Bounded Missing" section).

3.4 Predicted Mean Imputation

Imputed values are predicted using an ordinary least squares (OLS) multiple regression algorithm to impute the most likely value when the variable to be imputed is continuous or ordinal. When the variable to be imputed is a binary or categorical variable, a discriminant method is applied to impute the most likely value.

3.5 Ordinary Least Squares

Using the least squares method, missing values are imputed using predicted values from the corresponding covariates using the estimated linear regression models. This method is used to impute all the continuous variables in a data set.

3.6 Discriminant

Discriminant multiple imputation is a model-based method for binary or categorical variables.

3.7 Multiple Imputation Overview

SOLAS™ 3.0 provides two distinct methods for performing multiple imputation:

Predictive Model-Based Method.
Propensity Score.

3.8 Predictive Model-Based Method

The models that are available at present are ordinary least squares regression and a discriminant model. When the data are continuous or ordinal, the OLS method is applied. When the data are categorical, the discriminant method is applied.

Multiple imputations are generated using a regression model of the imputation variable on a set of user-specified covariates. The imputations are generated via randomly drawn regression model parameters from the Bayesian posterior distribution based on the cases for which the imputation variable is observed.

Each imputed value is the predicted value from these randomly drawn model parameters plus a randomly drawn error term. The randomly drawn error term is added to the imputations to prevent over-smoothing of the imputed data. The regression model parameters are drawn from a Bayesian posterior distribution to reflect the extra uncertainty due to the fact that the regression parameters can be estimated, but not determined, from the observed data.

3.9 Propensity Score Method

The system applies an implicit model approach based on propensity scores and an Approximate Bayesian Bootstrap to generate the imputations. The propensity score is the estimated probability that a particular element of data is missing. The missing data are filled in by sampling from the cases that have a similar propensity to be missing. The multiple imputations are independent repetitions from a Posterior Predictive Distribution for the missing data, given the observed data.

4 MULTIPLE IMPUTATION IN SOLAS™ 3.0

Multiple imputation replaces each missing value in the data set with several imputed values instead of just one. First proposed by Rubin in the early 1970s as a possible solution to the problem of survey nonresponse, the method corrects the major problems associated with single imputation. Multiple imputation creates M imputations for each missing value, thereby reflecting the uncertainty about which value to impute (Scheme A.1).

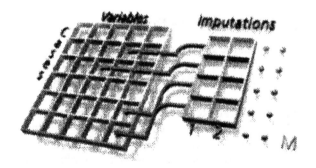

Scheme A.1

The first set of the M imputed values is used to form the first imputed data set, the second set of the M imputed values is used to form the second imputed data set, and so on. In this way, M imputed data sets are obtained. Each of the M imputed data sets is statistically analyzed by the complete-data method of choice. This yields M intermediate results. These M intermediate results are then combined into a final result, from which the conclusions are drawn, according to explicit formulas. The extra inferential uncertainty due to missing data can be assessed by examining the between-imputation variance and the following related measures:

> The relative increases in variance due to nonresponse (R_m) and the fraction of information missing due to missing data (Y_m).

5 GENERAL

Before the imputations are actually generated, the missing data pattern is sorted as close as possible to a monotone missing data pattern, and each missing data entry is either labeled as monotone missing or nonmonotone missing, according to where it fits in the sorted missing data pattern.

5.1 Monotone Missing Data Pattern

A monotone missing data pattern occurs when the variables can be ordered, from left to right, such that a variable to the left is at least as observed as all variables to the right. For example, if variable A is fully observed and variable B is sometimes missing, A and B form a monotone pattern. Or if A is only missing if B is also missing, A and B form a monotone pattern. If A is sometimes missing when B is observed, and when B is sometimes missing when A is observed, then the pattern is not monotone (e.g., see Little and Rubin, 1987).

We also distinguish between a missing data pattern and a local missing data pattern:

> A missing data pattern refers to the entire data set, such as a monotone missing data pattern.
> A local missing data pattern for a case refers to the missingness for a particular case of a data set.
> A local missing data pattern for a variable refers to the missingness for that variable.

If two cases have the same sets of observed variables and the same sets of missing variables, then these two cases have the same local missing data pattern.

A monotone pattern of missingness, or a close approximation to it, can be quite common. For example, in longitudinal studies, subjects often drop out as the study progresses so that all subjects have time 1 measurements, a subset of subjects has time 2 measurements, only a subset of those subjects has time 3 measurements, and so on. SOLAS™ sorts variables and cases into a pattern that is as close as possible to a monotone pattern. Monotone patterns are attractive because the resulting analysis is flexible and is completely principled because only observed/real data are used in the model to generate the imputed values.

6 PREDICTIVE MODEL-BASED METHOD

If Predictive Model-Based Multiple Imputation is selected, then an ordinary least squares regression method of imputation is applied to the continuous, integer, and ordinal imputation variables, and discriminant multiple imputation is applied to the nominal imputation variables.

6.1 Ordinary Least Squares Regression

The predictive information in a user-specified set of covariates is used to impute the missing values in the variables to be imputed. First, the Predictive Model is estimated from the observed data. Using this estimated model, new linear regression parameters are randomly drawn from their Bayesian posterior distribution. The randomly drawn values are used to generate the imputations, which include random deviations from the model's predictions. Drawing the exact model from its posterior distribution ensures that the extra uncertainty about the unknown true model is reflected.

In the system, multiple regression estimates of parameters are obtained using the method of least squares. If you have declared a variable to be nominal, then you need design variables (or dummy variables) to use this variable as a predictor variable in a multiple linear regression. The system's multiple regression allows for this possibility and will create design variables for you.

6.2 Generation of Imputations

Let Y be the variable to be imputed, and let X be the set of covariates. Let Y_{obs} be the observed values in Y, and let Y_{mis} be the missing values in Y. Let X_{obs} be the units corresponding to Y_{obs}.

The analysis is performed in two steps:

1. The Linear Regression-Based Method regresses Y_{obs} on X_{obs} to obtain a prediction equation of the form: $\hat{Y}_{mis} = a + bX_{mis}$.

2. A random element is then incorporated in the estimate of the missing values for each imputed data set. The computation of the random element is based on a posterior drawing of the regression coefficients and their residual variances.

6.3 Posterior Drawing of Regression Coefficients and Residual Variance

Parameter values for the regression model are drawn from their posterior distribution given the observed data using noninformative priors. In this way, the extra uncertainty due to the fact that the regression parameters can be estimated, but not determined, from Y_{obs} and X_{obs} is reflected.

Using estimated regression parameters rather than those drawn from its posterior distribution can produce poor results, in the sense that the between-imputation variance is underestimated.

6.4 Discriminant Multiple Imputation

Discriminant multiple imputation is a model-based method for imputing binary or categorical variables.

Let $i, \ldots, s$ be the categories of the categorical imputation variable y. Bayes theorem is used to calculate the probability that a missing value in the imputation variable y is equal to its jth category given the set of the observed values of the covariates and of y.

7 PROPENSITY SCORE

The system applies an implicit model approach based on propensity scores and an Approximate Bayesian Bootstrap to generate the imputations. The underlying assumption about Propensity Score Multiple Imputation is that the nonresponse of an imputation variable can be explained by a set of covariates using a logistic regression model. The multiple imputations are independent repetitions from a Posterior Predictive Distribution for the missing data, given the observed data.

Variables are imputed from left to right through the data set, so that values that are imputed for one variable can be used in the prediction model for missing values occurring in variables to the right of it. The system creates a temporary variable that will be used as the dependent variable in a logistic regression model. This temporary variable is a response indicator and will equal 0 for every case in the imputation variable that is missing and will equal 1 otherwise.

The independent variables for the model will be a set of baseline/fixed covariates that we believe are related to the variable we are imputing. For example, if the variable being imputed is period t of a longitudinal variable, the covariates might include the previous periods $(t-1, t-2, \ldots, t-n)$.

The regression model will allow us to model the "missingness" using the observed data. Using the regression coefficients, we calculate the propensity that a subject would have a missing value in the variable in question. In other words, the propensity score is the conditional probability of "missingness," given the vector of observed covariates. Each missing data entry of the imputation variable y is imputed by values randomly drawn from a subset of observed values of y (i.e., its donor pool), with an assigned probability close to the missing data entry that is to be imputed. The donor pool defines a set of cases with observed values for that imputation variable.

Covariates that are used for the generation of the imputations are selected for each imputation variable separately. For each imputation variable, two sets of covariates are selected. One set of covariates is used for imputing the nonmonotone missing data entries and the other set of covariates is used for imputing the monotone missing data entries in that variable. Which missing data entries are labeled as nonmonotone or monotone is determined after the missing data pattern has been sorted. For both sets of selected covariates for an imputation variable, a special subset is the fixed covariates.

Fixed covariates are all selected covariates other than imputation variables, and are used for the imputation of missing data entries for monotone and nonmonotone missing patterns. This is only the case for fixed covariates.

7.1 Defining Donor Pools Based on Propensity Scores

Using the options in the Donor Pool window, the cases of the data sets can be partitioned into c donor pools of respondents according to the assigned propensity scores, where $c = 5$ is the default value of c. This is done by sorting the cases of the data sets according to their assigned propensity scores in ascending order.

The Donor Pool page gives the user more control over the random draw step in the analysis. You are able to set the subclass ranges and refine these ranges further using another variable, known as the refinement variable, described below. Three ways of defining the donor pool sub-classes are provided:

1. You can divide the sample into c equal-sized subsets; the default will be five. If the value of c results in not more than one case being available to the selection algorithm, c will decrement by one until such time as there are sufficient data. The final value of c used is included in the imputation report output described later in this manual.

2. You can use the subset of c cases that are closest with respect to propensity score. This option allows you to specify the number of cases before and after the case being imputed, which are to be included in the subclass. About 50% of the cases will be used before, and 50% of the cases will be used after. The default c will be 10 and cannot be set to a value less than two. If less than two cases are available, a value of five will be used for c.

3. You can use the subset of d% of the cases that are closest with respect to propensity score. This option allows you to specify the number of cases before and after the case being imputed. This is the percentage of "closest" cases in the data set to be included in the subclass. The default for d will be 10.00, and cannot be set to a value that will result in less than two cases being available. If less than two cases are available, a d value of five will be used.

7.2 Refinement Variable

Using the Donor Pool window, a refinement variable w can be chosen. For each missing value of y that is to be imputed, a smaller subset is selected on the basis of the association between y and w. This smaller subset will then be used to generate the imputations.

For each missing value of y, the imputations are randomly drawn according to the Approximate Bayesian Bootstrap method from the chosen subset of observed values of y. Using this method, a random sample (with replacement) is randomly drawn from the chosen subset of observed values to be equal in size to the number of observed values in this subset. The imputations are then randomly drawn from this sample.

The Approximate Bayesian Bootstrap method is applied in order to reflect the extra uncertainty about the predictive distribution of the missing value of y, given the chosen subset of observed values of y. This predictive distribution can be estimated from the chosen subset of observed values of y, but not determined. By drawing the imputations randomly from the chosen subset of observed values rather than applying the Approximate Bayesian Bootstrap, this results in improper imputation in the sense that the between-imputation variance is underestimated.

7.3 Bounded Missing

This type of missing value can only occur when a variable is longitudinal. It is a missing value that has at least one observed value before, and at least one observed value after, the period for which it is missing. The following table shows an example of bounded missing values. The variables Month 1 to Month 6 are a set of longitudinal measures:

Patient	Month 1	Month 2	Month 3	Month 4	Month 5	Month 6
101	10	*	*	20	*	50
102	20	40	*	30	*	*
103	30	*	*	*	*	50

* = missing, shaded = bounded missing

    Linear interpolation can be used to fill in missing values that are longitudinal variables. So, for example, using linear interpolation, Patient 101's missing values for Months 2, 3, and 5 would be imputed as follows (Scheme A.2):

    Thus the imputed value for Month 2 will be 13.33, the imputed value for Month 3 will be 16.67, and the imputed value for Month 5 will be 35.

8 GENERATING MULTIPLE IMPUTATIONS

Once the missing data pattern has been sorted and the missing data entries have been labeled either as nonmonotone missing or monotone missing, the imputations are generated in two steps:

    1. The nonmonotone missing data entries are imputed first.
    2. Then the monotone missing data entries are imputed using the previously imputed data for the nonmonotone missing data entries.

    The nonmonotone missing data entries are always imputed using a Predictive Model-Based Multiple Imputation. The monotone missing

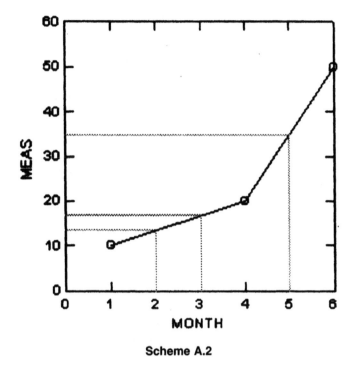

Scheme A.2

data entries are imputed by the user-specified method, which can be either the Propensity Score Method or the Predictive Model-Based Method.

8.1 Imputing the Nonmonotone Missing Data

The Nonmonotone missing data are imputed for each subset of missing data by a series of individual linear regression multiple imputations (or discriminant multiple imputations) using, as much as possible, observed and previously imputed data.

First, the leftmost nonmonotone missing data are imputed. Then the second leftmost nonmonotone missing data are imputed using the previously imputed values. This is continued, until the rightmost nonmonotone missing data are imputed using the previously imputed values for the other nonmonotone missing data in the same subset of cases.

The user can specify or add covariates for use in the Predictive Model for any variables that will be imputed. More information about using covariates is given in the example below.

8.2 Imputing the Monotone Missing Data

The monotone missing data are sequentially imputed for each set of imputation variables with the same local pattern of missing data. First, the leftmost set is imputed using the observed values of this set and its selected fixed covariates only. Then the next set is imputed using the observed values of this set, the observed and previous imputed values of the first set, and the selected fixed covariates.

This continues until the monotone missing data of the last set are imputed. For each set, the observed values of this set, the observed and imputed values of the previously imputed sets, and the fixed covariates are used. If multivariate Propensity Score Multiple Imputation is selected for the imputation of the monotone missing data, then this method is applied for each subset of sets having the same local missing data pattern.

9 SHORT EXAMPLES

These short examples use the data set **MI_TRIAL.MDD** (located in the **SAMPLES** subdirectory). This data set contains the following 11 variables measured for 50 patients in a clinical trial:

> **OBS**—observation number.
> **SYMPDUR**—duration of symptoms.
> **AGE**—patient's age.
> **MeasA_0, MeasA_1, MeasA_2,** and **MeasA_3**—baseline measurement for the response variable MeasA and three postbaseline measurements taken at Months 1–3.
> **MeasB_0, MeasB_1, MeasB_2,** and **MeasB_3**—baseline measurement for the response variable MeasB and three postbaseline measurements taken at Months 1–3.

The variables **OBS, SYMPDUR, AGE, MeasA_0,** and **MeasB_0** are all fully observed, and the remaining six variables contain missing values. To view the missing pattern for this data set, do the following:

1. From the datasheet window, select **View** and **Missing Pattern**. In the Specify Missing Data Pattern window, press the **Use All** button.
2. From the **View** menu of the Missing Data Pattern window, select **View Monotone Pattern** to display the window shown below left

(Scheme A.3). Note that after sorting the data into a monotone pattern, the time structure of the longitudinal measures is preserved, so the missing data pattern in this data set is monotone over time.

3. To close the **Missing Data Pattern window**, select **File** and **Close**.

9.1 Predictive Model-Based Method—Example

We will now multiply impute all of the missing values in this data set using the Predictive Model-Based Method by executing the following steps:

1. From the **Analyze** menu, select **Multiple Imputation** and **Predictive Model-Based Method**.

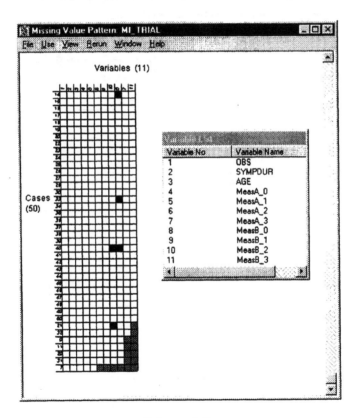

Scheme A.3

2. The Specify Predictive Model window is displayed. The window opens with two pages or tabs: **Base Setup** and **Advanced Options**. As soon as you select a variable to be imputed, a **Nonmonotone** tab and a **Monotone** tab are also displayed.

9.2 Base Setup

Selecting the **Base Setup** tab allows you specify which variables you want to impute, and which variables you want to use as covariates for the predictive model (Scheme A.4):

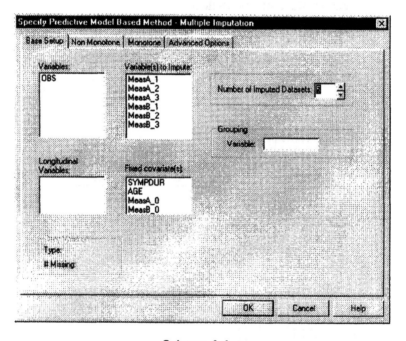

Scheme A.4

1. Drag-and-drop the variables **MeasA_1, MeasA_2, MeasA_3, MeasB_1, MeasB_2,** and **MeasB_3** into the Variables to Impute field.
2. Drag-and-drop the variables **SYMPDUR, AGE, MeasA_0**, and **MeasB_0** into the Fixed Covariates field.

3. Because there is no grouping variable in this data set, we can leave this field blank.

9.3 Nonmonotone

Selecting the **Nonmonotone** tab allows you to add or remove covariates from the predictive model used for imputing the nonmonotone missing values in the data set. (These can be identified in the **Missing Data Pattern** mentioned earlier.) You select the + or − sign to expand or contract the list of covariates for each imputation variable (Scheme A.5).

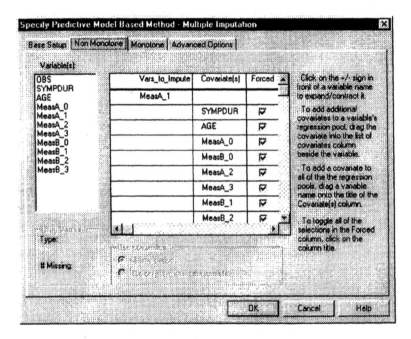

Scheme A.5

For each imputation variable, the list of covariates will be made up of the variables specified as Fixed Covariates in the **Base Setup** tab, and all of the other imputation variables. Variables can be added and removed from this list of covariates by simply dragging and dropping the variable from the covariate list to the variables field, or vice versa. Even though a variable appears in the list of covariates for a particular imputation variable, it might not be used in the final model.

The program first sorts the variables so that the missing data pattern is as close as possible to monotone, and then, for each missing value in the imputation variable, the program works out which variables, from the total list of covariates, can be used for prediction.

By default, all of the covariates are forced into the model. If you uncheck a covariate, it will not be forced into the model, but will be retained as a possible covariate in the stepwise selection. Details of the models that were actually used to impute the missing values are included in the **Regression Output**, which can be selected from the **View** menu of the Multiply Imputed Data Pages. These data pages will be displayed after you have specified the imputation and pressed the **OK** button in the Specify Predictive Model window.

9.4 Monotone

Selecting the **Monotone** tab allows you to add or remove covariates from the predictive model used for imputing the monotone missing values in the data set. (These can be identified in the **Missing Data Pattern** mentioned earlier.) Again, you select the + or − sign to expand or contract the list of covariates for each imputation variable (Scheme A.6).

For each imputation variable, the list of covariates will be made up of the variables specified as Fixed Covariates in the **Base Setup** tab, and all of the other imputation variables. Variables can be added and removed from this list by simply dragging and dropping. Even though a variable appears in the list of covariates for a particular imputation variable, it might not be used in the final model. The program first sorts the variables so that the missing data pattern is as close as possible to monotone, and then uses only the variables that are to the left of the imputation variable as covariates. Details of the models that were actually used to impute the missing values are included in the **Regression Output**.

9.5 Advanced Options

Selecting the **Advanced Options** tab displays a window that allows you to choose control settings for the regression/discriminant mode (Scheme A.7).

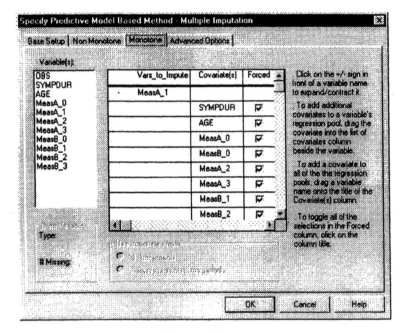

Scheme A.6

9.6 Tolerance

The value set in the **Tolerance** data field controls numerical accuracy. The tolerance limit is used for matrix inversion to guard against singularity. No independent variable is used whose R^2 with other independent variables exceeds (1−Tolerance). You can adjust the tolerance using the scrolled data field.

9.7 Stepping Criteria

Here you can select **F-to-Enter** and **F-to-Remove** values from the scrolled data fields, or enter your chosen value. If you wish to see more variables entered in the model, set the **F-to-Enter** value to a smaller value. The numerical value of **F-to-Remove** should be chosen to be less than the **F-to-Enter** value.

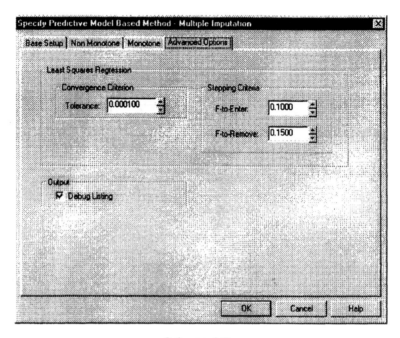

Scheme A.7

When you are satisfied that you have specified your analysis correctly, click the **OK** button. The multiply imputed data pages will be displayed, with the imputed values appearing in blue. Refer to the sections "Multiple Imputation Output" and "Analyzing Multiply Imputed Data Sets—Example" for further details about analyzing these data sets and combining the results.

9.8 Propensity Score Method—Example

We will now multiply impute all of the missing values in the data set using the Propensity Score-Based Method:

1. From the **Analyze** menu, select **Multiple Imputation** and **Propensity Score Method**.
2. The **Specify Propensity Method** window is displayed and is a tabbed (paged) window. The window opens with two pages or tabs: **Base Setup** and **Advanced Options**. As soon as you select a

variable to be imputed, a **Nonmonotone** tab, a Monotone tab, and a **Donor Pool** tab are also displayed.

9.9 Base Setup

Selecting the **Base Setup** tab allows you specify which variables you want to impute, and which variables you want to use as covariates for the logistic regression used to model the missingness (Scheme A.8):

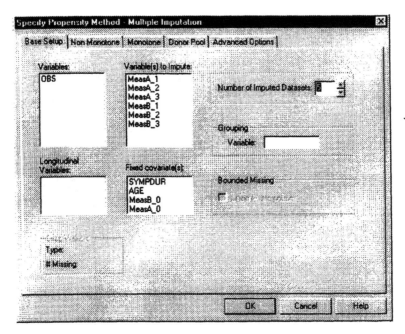

Scheme A.8

1. Drag-and-drop the variables **MeasA_1**, **MeasA_2**, **MeasA_3**, **MeasB_1**, **MeasB_2**, and **MeasB_3** into the Variables to Impute field.
2. Drag-and-drop the variables **SYMPDUR**, **AGE**, **MeasA_0**, and **MeasB_0** into the Fixed Covariates field.
3. As there is no grouping variable in this data set, we can leave this field blank.

9.10 Nonmonotone

Selecting the **Nonmonotone** tab allows you to add or remove covariates
from the logistic model used for imputing the nonmonotone missing values
in the data set. (These can be identified in the **Missing Data Pattern**
mentioned earlier in the Predictive Model example.)

You select the + or − sign to expand or contract the list of covariates
for each imputation variable (Scheme A.9).

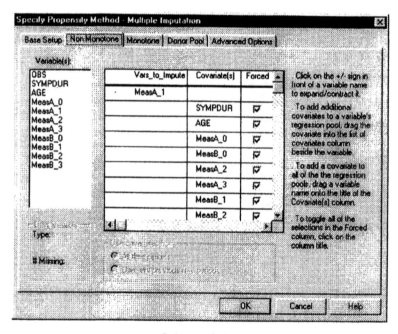

Scheme A.9

The list of covariates for each imputation variable will be made up of
the variables specified as Fixed Covariates in the **Base Setup** tab, and all of
the other imputation variables. Variables can be added and removed from
this list of covariates by simply dragging and dropping the variable from
the covariate list to the variables field, or vice versa. Even though a variable
appears in the list of covariates for a particular imputation variable, it
might not be used in the final model.

The program first sorts the variables so that the missing data pattern is as close as possible to monotone, and then, for each missing value in the imputation variable, the program works out which variables, from the total list of covariates, can be used for prediction.

By default, all of the covariates are forced into the model. If you uncheck a covariate, it will not be forced into the model, but will be retained as a possible covariate in the stepwise selection. Details of the models that were actually used to impute the missing values are included in the **Regression Output**, which can be selected from the **View** menu of the Multiply Imputed Data Pages. These data pages will be displayed after you have specified the imputation and pressed the **OK** button in the Specify Predictive Model window.

9.11 Monotone

Selecting the **Monotone** tab allows you to add or remove covariates from the logistic model used for imputing the monotone missing values in the data set. (These can be identified in the **Missing Data Pattern** mentioned earlier) (Scheme A.10).

Again, you select the + or − sign to expand or contract the list of covariates for each imputation variable.

The list of covariates for each imputation variable will be made up of the variables specified as Fixed Covariates in the **Base Setup** tab, and all of the other imputation variables. Variables can be added and removed from this list by simply dragging and dropping the variable from the list of covariates to the variables field, or vice versa. Even though a variable appears in the list of covariates for a particular imputation variable, it might not be used in the final model.

The program first sorts the variables so that the missing data pattern is as close as possible to monotone, and then uses only the variables that are to the left of the imputation variable as covariates. Details of the models that were actually used to impute the missing values are included in the **Regression Output**.

9.12 Donor Pool

Selecting the **Donor Pool** tab displays the Donor Pool page that allows more control over the random draw step in the analysis by allowing the user to define propensity score subclasses (Scheme A.11).

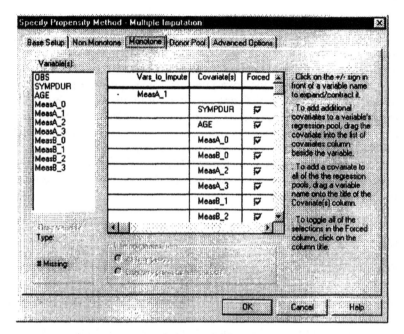

Scheme A.10

The following options for defining the propensity score subclasses are provided:

Divide propensity score into *c* subsets. The default is 5.

Use *c* closest cases. This option allows you to specify the number of cases, before and after the case, being imputed that are to be included in the subset.

Use *d*% of the data set closest cases. This option allows you to specify the number of cases as a percentage.

See "Defining Donor Pools Based on Propensity Scores" section.

You can use one refinement variable for each of the variables being imputed. Variables can be dragged from the **Variables** listbox to the **Refinement Variable** column. When you use a refinement variable, the program reduces the subset of cases included in the donor pool to include only cases that are close with respect to their values of the refinement variable.

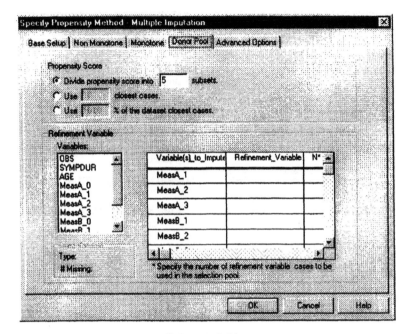

Scheme A.11

You can also specify the number of refinement variable cases to be used in the donor pool. For this example, we will use all of the default settings in this tab.

9.13 Advanced Options

Selecting the **Advanced Options** tab displays the Advanced Options window that allows the user to control the settings for the imputation and the logistic regression (Scheme A.12).

9.14 Randomization

The **Main Seed Value** is used to perform the random selection within the propensity subsets. The default seed is 12345. If you set this field to blank, or set it to zero, then the clock time is used.

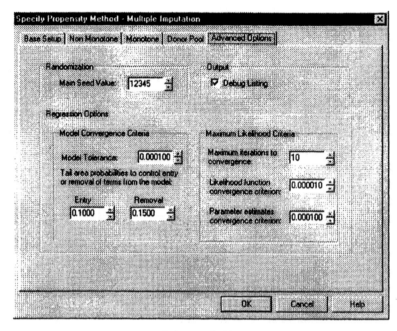

Scheme A.12

9.15 Regression Options

The value set in the **Tolerance** data field controls numerical accuracy. The tolerance limit is used for matrix inversion to guard against singularity. No independent variable is used whose R^2 with other independent variables exceeds $(1-\text{Tolerance})$. You can adjust the tolerance using the scrolled data field.

9.16 Stepping Criteria

Here you can select **F-to-Enter** and **F-to-Remove** values from the scrolled data fields, or enter your chosen value. If you wish to see more variables entered in the model, set the **F-to-Enter** value to a smaller value. The numerical value of **F-to-Remove** should be chosen to be less than the **F-to-Enter** value.

When you are satisfied that you have specified your analysis correctly, click the OK button. The multiply imputed data pages will be displayed, with the imputed values appearing in blue. Refer to "Analyzing

Multiply Imputed Data Sets" for further details of analyzing these data sets and combining the results.

9.17 Multiple Imputation Output

The multiple imputation output, either Propensity Score or the Predictive Model-Based Method, comprises five (default value) Multiple Imputation Data Pages. From the View menu of the Data Pages, you can select either **Imputation Report, Regression Output,** or **Missing Pattern.**

When other analyses are performed from the **Analyze** menu of a data page (see the section "Analyzing Multiply Imputed Data Sets—Example"), a **Combined** tab is added to the data page tabs. Selecting this tab displays the combined statistics for these data pages.

9.18 Data Pages

The multiple imputation output displays five data pages with the imputed values shown in blue (two of the values are shown here highlighted). The first data page (page 1) for the above example is shown below (Scheme A.13).

From the **View** menu, you can select **Imputation Report** and **Regression Output** (examples of both are shown below) or **Missing Pattern** (Scheme A.14).

The Imputation Report and a Regression Output (shown in part above) summarize the results of the logistic regression, the ordinary regression, and the settings used for the multiple imputation.

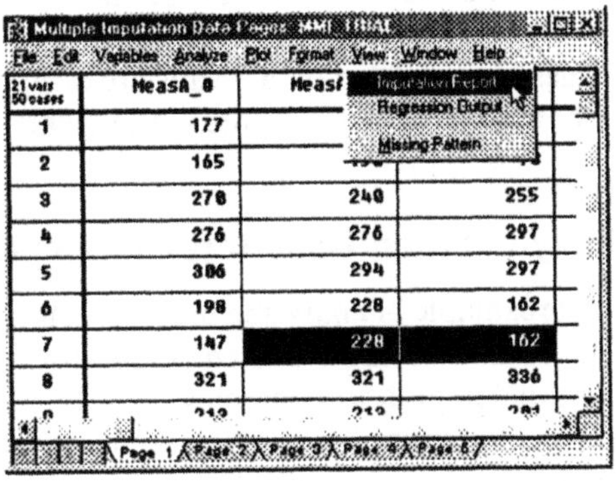

Scheme A.13

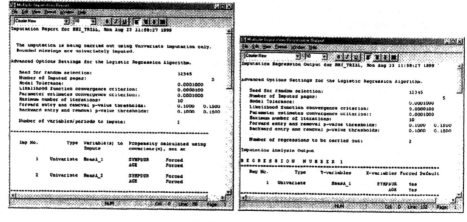

Scheme A.14

9.19 Multiple Imputation Report

The imputation report contains a summary of the parameters that were chosen for the multiple imputation. For example, the seed value that was used for the random selection, the number of imputations that were performed, etc., are all reported. The report shows:

An overview of the multiple imputation parameters.
The equations used to generate the imputations.
Additional diagnostical information that can be used to judge the quality and validity of the generated imputations.

Conclusions about the statistical analysis can be drawn from the combined results (see "Analyzing Multiply Imputed Data Sets—Example" section). These five pages of completed data results are displayed and allow the user to examine how the combined results are calculated.

9.20 Analyzing Multiply Imputed Data Sets—Example

This section presents a simple example of analyzing multiply imputed data sets. It will show how the results of the repeated imputations can be combined to create one repeated imputation inference.

After you have performed a multiple imputation on your data set, you will have M complete data sets, each of which can be analyzed using standard complete-data statistical methods.

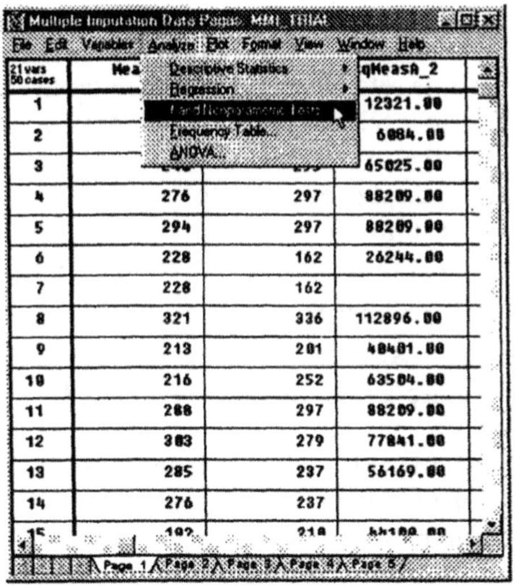

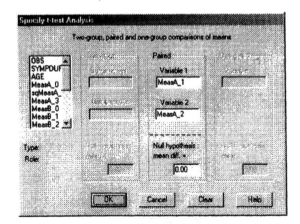

Scheme A.15

Scheme A.16

If you select **Descriptive Statistics**, **Regression**, *t*-**Test**, **Frequency Table**, or **ANOVA** from the Analyze menu from any data page, the analysis will be performed on all five data sets. The analysis generates five pages of output, one corresponding to each of the imputed data sets, and a Combined page which gives the overall set of results. The tabs at the bottom of a page allow you to display each data set.

This example uses the imputation results from the data set **MI_TRIAL.MDD** that was used in the propensity score example earlier. Part of data in page 1 for that example is shown below (Scheme A.15):

1. From the data page **Analyze** menu, select *t*-**Test and Non-parametric Test** to display the Specify *t*-Test Analysis window (Scheme A.16).
2. Drag-and-drop the variables **MeasA_1** and **MeasA_2** to the **Variable 1** and **Variable 2** data fields, respectively.
3. Press the **OK** button to display the data pages, then the **Combine** tab to display the combined statistics from the five imputed data pages (Scheme A.17).

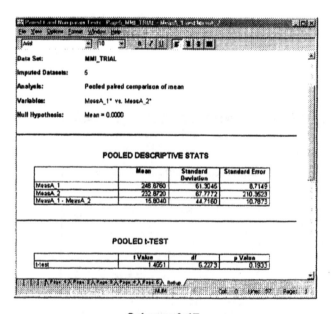

Scheme A.17

Appendix B: ℓ_{EM}

ℓ_{EM} is a software for analyzing categorical data, developed by Jeroen K. Vermunt at Tilburg University. The name ℓ_{EM} stems from "log-linear and event history analysis with missing data using the EM algorithm" (Vermunt, 1997). As the name suggests, the software is for analysis based on models expressed as a log-linear model. On the Internet, the software is publicized at several websites as a freeware. For example, refer to the website on latent class analysis software (http://members.xoom.com/jsuebersax/soft.html).

Two versions of ℓ_{EM} are available: DOS and Windows versions, both of which require the creation and execution of an input file according to the syntax of ℓ_{EM} (refer to the next paragraph). The input file requires a description of data, model, estimation options, output options, etc.

In the Windows version, three windows appear on the screen, one each for input, log, and output. In the input window, the data subject to analysis and the analysis model must be described. If a file that has already been created is read in, this file will be displayed in that window. The log window records the log when the commands described in the input window are executed. This is used when there is an error in the description of the model and correction is required. The output window displays the results of executing the commands in the input window. The output can be changed by using pre-prepared commands and options (Scheme B.1).

ℓ_{EM} consists of the software program, a 100-page manual and 229 example files. As the manual is a PostScript file, it can be viewed on the computer screen and printed out from a PostScript file browsing software such as GSView (Scheme B.2).

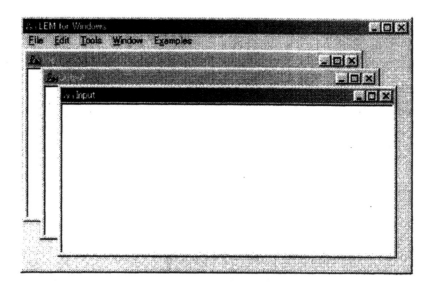

Scheme B.1

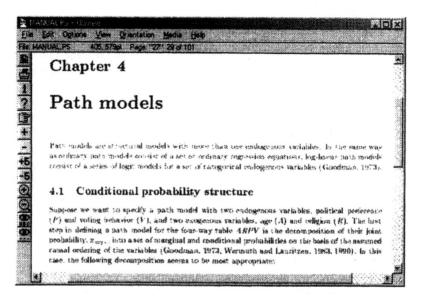

Scheme B.2

The example files can be directly read in as input files. However, it is easier to select [Examples] on the menu bar in the startup screen to display a tree menu, which is more convenient.

The example shown in the next figure reads in an example of a latent class model. The latent class analysis reviewed in Chap. 5 explains the relationship among manifest variables by using latent variables assuming category values, and can be regarded as an analysis based on a log-linear model in which there is one latent variable (Scheme B.3).

After reading in the input file, select the [Run] option from [File] on the menu bar. Calculations to estimate the parameter included in the model will automatically be executed, and various statistics will be outputted (Scheme B.4).

As the calculation results are displayed in the output window, activate the output window to view them (Scheme B.5).

All estimates of the parameters outputted by ℓ_{EM} are maximum likelihood estimates. Generally, maximum likelihood estimates of parameters of hierarchical log-linear models are determined with the use of an iterative proportional fitting (IPF) algorithm. In cases where latent variables are included, the latent variables are regarded as missing data and the EM algorithm is used. In E-step, the conditional expected value of the log likelihood of the complete data including latent variables is calculated, whereas in M-step, the parameter value is updated with the use of the IPF algorithm, etc. ℓ_{EM} uses the ECM algorithm in many cases.

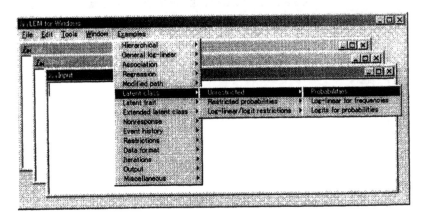

Scheme B.3

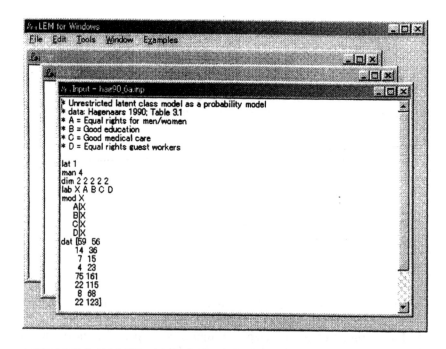

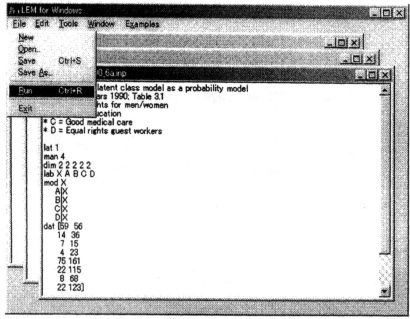

Scheme B.4

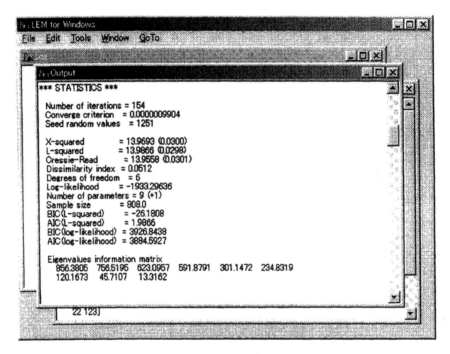

Scheme B.5

A latent class model is expressed as a log-linear model with one latent variable. An extension of this model is a log-linear model with more than one latent variable. ℓ_{EM} can flexibly handle such models as well. Refer to Hagenaars (1993) for analysis based on such models.

REFERENCES

1. Hagenaars, J. A. (1993). *Loglinear Models with Latent Variables*. Newbury Park, CA: Sage.
2. Vermunt, J. K. (1997). *Lem: A General Program for the Analysis of Categorical Data*. Tilburg, Netherlands: Tilburg University.

Index